Troubled Geographies

THE SPATIAL HUMANITIES   David J. Bodenhamer, John Corrigan,
and Trevor M. Harris, editors

TROUBLED GEOGRAPHIES

A SPATIAL HISTORY OF RELIGION AND SOCIETY IN IRELAND

IAN N. GREGORY, NIALL A. CUNNINGHAM,
C. D. LLOYD, IAN G. SHUTTLEWORTH,
AND PAUL S. ELL

INDIANA UNIVERSITY PRESS   Bloomington & Indianapolis

This book is a publication of

Indiana University Press
Office of Scholarly Publishing
Herman B Wells Library 350
1320 East 10th Street
Bloomington, Indiana 47405 USA

Telephone orders 800-842-6796
Fax orders 812-855-7931

⊖ The paper used in this publication meets the minimum requirements of the American National Standard for Information Sciences— Permanence of Paper for Printed Library Materials, ANSI Z39.481992.

Manufactured in China

Library of Congress Cataloging-in-Publication Data

Gregory, Ian N.
  Troubled geographies : a spatial history of religion and society in Ireland / Ian N. Gregory, Niall A. Cunningham, Paul S. Ell, C. D. Lloyd, and Ian G. Shuttleworth.
    pages cm. — (The spatial humanities)
  Includes bibliographical references and index.
  ISBN 978-0-253-00966-1 (cloth : alk. paper) — ISBN 978-0-253-00973-9 (pbk. : alk. paper) — ISBN (invalid) 978-0-253-00979-1 (ebook) 1. Human geography—Ireland. 2. Ireland—Ethnic relations. 3. Ireland—Religious life and customs. 4. Ireland—Social life and customs. I. Title.
  GF563.G74 2013
  304.209415—dc23
                              2013016440

1 2 3 4 5 18 17 16 15 14 13

Contents

Figures

F

Tables

T

Acknowledgments

The research that produced this book was funded by the Arts and Humanities Research Council/Economic and Social Research Council's Religion and Society Programme under grant AH/F008929/1 "Troubled Geographies: Two Centuries of Religious Division in Ireland." It benefited from additional support from the British Academy under grant SG090803 "The Presbyterian Church in Ireland, 1871–2001." Our sincere thanks go to Malcolm Sutton for allowing us to use his database of deaths during the Troubles and to Martin Melaugh (University of Ulster) for making the data set available to us. Many of the census data used have been taken from L. A. Clarkson, L. Kennedy, E. M. Crawford, and M. E. Dowling, *Database of Irish Historical Statistics, 1861–1911* (computer file) (Colchester, Essex: UK Data Archive [distributor], November 1997, study number 3579); and from M. W. Dowling, L. A. Clarkson, L. Kennedy, and E. M. Crawford, *Database of Irish Historical Statistics: Census Material, 1901–1971* (computer file) (Colchester, Essex: UK Data Archive [distributor], May 1998, study number 3542). Our thanks go also to Elaine Yeates and others at the Centre for Data Digitisation and Analysis, Queen's University Belfast, for additional work on these and related data sets. Census data for the Republic of Ireland for 1981 to 2002 were provided by the Central Statistics Office, Ireland. Access to the grid square data for Northern Ireland were provided by the Census Office of the Northern Ireland Statistics and Research Agency. Their use in this analysis benefited from Economic and Social Research Council grant RES-000-23-0478. Spatial data on Belfast's peace lines were downloaded from the Northern Ireland Statistics and Research Office website, http://www.nisra.gov.uk/geography/default.asp12.htm (23 September 2011). Many of the color schemes used benefited from input from ColorBrewer.org, http://www.colorbrewer.org (23 September 2011).

Troubled Geographies

Geography, Religion, and Society in Ireland: A Spatial History

Even today, more than a decade after the Belfast or Good Friday Agreement, which marked an end to the Troubles, the visitor to Northern Ireland cannot help but be struck by the interplay between religion, ethnonational identity, politics, history, and geography. Protestant areas are demarked by the Union Flag (the flag of the United Kingdom of Great Britain and Northern Ireland, formerly the United Kingdom of Great Britain and Ireland), backed up by red, white, and blue curbstones and murals representing events such as the Battle of the Boyne and the Siege of Derry. Protestantism is seen as synonymous with the politics of unionism and loyalism, which have the union with Great Britain and loyalty to the British Crown as their core tenets. Orange parades further emphasize these links—Orangemen march to church in a symbolic way that makes explicit the links between their religion, politics, history, and, most controversially, territory. So too in Catholic areas, except the flags are those of the Republic of Ireland, the curbstones are green, white, and orange, and the murals tend to focus on the sufferings and tribulations of the Gaelic Irish population from the Norman Conquest all the way through to the recent Troubles.[1] Catholicism is seen as synonymous with Irish nationalism and republicanism, which have sought to remove British influence from Ireland.

Religion and territory thus are explicitly linked, a link that has at worst led to killing, arson, and other forms of violence aimed at establishing or protecting territorial control. While these are the most overt and unpleasant expressions of the impact of religious geography on Ireland, spatioreligious processes—the way in which religion and geography become intertwined with each other and a range of broader factors within society—have a long tradition. Recent work by Alexandra Walsham has drawn attention to this link, seeing religious space in Ireland as a sort of theological palimpsest, constantly being written and overwritten by competing Catholic and Protestant imaginings of the past.[2] The idea of a more substantive link between geography and religion underpins the political ideologies of nationalism and unionism, which have shaped the island of Ireland since the Famine.[3] The ways in which religion, society, and geography have evolved to shape Ireland over the past two centuries are the major themes that this book will explore in detail. It is important, however, to establish what we mean by religion. In this context it does not refer to religious practice, including acts of worship, church attendance, and systems of belief; instead, it is primarily concerned with religious identity, a person's background and the community with which that person identifies. As the opening paragraph makes clear, religion is often tied up with a wide variety of other factors in society, particularly ethnonational identity and politics, but also economic opportunity and a range of other issues affecting almost every aspect of a person's life.

This intermingling of geography, religion, and the wider society is not new. Protestantism arrived in Ireland in the sixteenth and seventeenth centuries with the *plantations*, whose aim was to "plant" specific areas with English Protestants, loyal to the Crown, to defend English influence from the threats posed by indigenous Catholics. This took place against the backdrop of the wider European struggle between Protestant England and Catholic Spain and France. The plantations were focused on certain key strategic areas, including colonial Dublin and its historic sphere of influence known as the Pale, parts of Munster and the midlands, and west Ulster. Around the same time, Scottish Presbyterians arrived in large numbers in east Ulster, reflecting long-term economic and cultural links between southern Scotland and Ulster rather than the processes of large-scale, organized colonization.

The plantations laid the foundation for the fusing of religion, identity, politics, and geography, and, as chapter 2 will describe, many of the spatioreligious patterns that were laid down in this period have shaped, and been reshaped by, the processes that ran through nineteenth- and twentieth-century Ireland. These processes can be divided into two types. At one extreme there have been the short-term shocks—periods of intense violence or trauma that led to sudden upheavals. At the other there are the longer-term, more gradual processes associated with economic and social development.

The most obvious trauma was the period of violence from the Easter Rising in 1916 to Partition and the civil war in the early 1920s. The geographical legacy of this was the division of the island into the mainly Catholic Republic of Ireland—or the Irish Free State, as it was first called—and the predominantly Protestant Northern Ireland. As chapter 6 will identify, in this period religion, politics, and identity came together to tear the island in two. However, the resulting formal division of the island left both parts with significant populations of the "other" religion, and in some places in the north these minorities made up the majority of the population. In modern Northern Ireland this has had tragic consequences. The twentieth century's second period of violent trauma, described in the last chapters of the book, was the three decades of the Troubles, which have left Northern Ireland's two communities—defined by both religion and politics—as separated as they have ever been, emotionally as well as geographically. The so-called peace lines are the most obvious physical legacy of this period. These are high, ugly, concrete walls that scar Belfast's urban geography in their attempt to keep the two communities apart—a physical manifestation of a much deeper divide.[4] In what is now the Republic, however, there have been fewer such problems, and the much smaller Protestant minority has generally coexisted in relative harmony with their Catholic neighbors.

These two periods of violence are not the only traumatic processes to have affected the geographies of Ireland over the past two centuries. The Great Famine of the late 1840s was the last major famine in western Europe, and its effects have been profound in both the short and long term. Chapter 3 describes how, prior to the Famine, Ireland's population was increasing rapidly, as was typical of most nineteenth-century western countries. Chapter 4 moves on to describe how the Famine caused the population to crash from over eight million to less than six million as a consequence of death and emigration. Stagnation followed such that the population recorded by the 1841 census is still the largest ever population of Ireland, a situation that is virtually unique in western countries, where populations have typically increased dramatically since the early nineteenth century. Geography, religion, and politics are also part of the story of the Famine. It was perceived to have disproportionately affected Catholic areas in the west of the island, and the British government's response, or lack of it, to the unfolding tragedy, combined with the indifference of absentee landlords, did much to stimulate movements for agrarian reform and Home Rule.

As well as these shocks, Ireland has been affected by the more gradual but nevertheless highly significant processes that affected other western European countries over these two centuries as the island developed from an agrarian society to a postindustrial one. Urbanization, industrialization, suburbanization, and deindustrialization have all played their parts in shaping modern Ireland. Again, these processes have had marked spatial patterns and have had impacts on—and been impacted by—religion, identity, and politics in ways that are distinct from other parts of Europe. Chapter 5 describes how, during the nineteenth century, Belfast, located in the Protestant heartland of northeast Ulster, grew from almost nothing to become the largest city on the island. Its economy, based on shipbuilding, textiles, and other manufacturing industries, was firmly tied into the economy of Britain and the wider British Empire. Much of the rest of the island did not industrialize or urbanize. As a result, rural population pressures could not be absorbed by rapidly growing Irish cities and were instead absorbed by the cities of Britain and North America. The extent to which these trends have been shaped by religion is, at best, controversial, although the importance of the "Protestant work ethic" has been argued for.[5] These trends have, however, undoubtedly shaped religious geographies and with them a host of related themes. The rapid growth of Belfast led to a large influx of both Catholic and Protestant migrants into the city, leading to its complex sectarian geography. Economic marginalization of Catholics and discrimination against them in the shipyards in particular helped to foster resentments. The linkage of Belfast into the wider British economy was important in promoting unionism and the opposition to Home Rule, particularly among the Protestant elite.

More generally, as identified in both chapters 5 and 7, long-term processes have resulted in the population becoming increasingly concentrated in the towns and cities of the east coast at the expense of the west and center of the island. This has led to Catholics and Protestants living in closer proximity, which has sometimes, but not always, led to conflict. For much of the twentieth century the economic success of Northern Ireland and the stagnation of the Free State/Republic's economy exacerbated this conflict, with Belfast being economically and demographically the dominant center for much of the nineteenth and twentieth centuries.

The later twentieth century, however, saw a dramatic transition. As described in chapter 8, in the last decades of the twentieth century the Republic moved to open its economy and went through a period of rapid industrialization, followed by an even more rapid and spectacularly successful move to a service-based economy. By the end of the twentieth century the Celtic Tiger had become the fourth richest country in the world.[6] Over the same period, as chapters 9 to 11 describe, Northern Ireland's traditional manufacturing industries went into steep decline. Efforts to replace them were undermined by the Troubles, the conflict between Catholics and Protestants that erupted in the late 1960s. The resulting violence not only was a human tragedy but did much to undermine Northern Ireland's attempts to reinvent itself as a postindustrial society, meaning that by the end of the twentieth century the Republic had the strong economy, while the north stagnated and declined.

This book is thus concerned with how a geography laid down in the sixteenth and seventeenth centuries led to an amalgamation of religious conviction, ethnonational identity, and political opinions that shaped the geographies of Ireland through the nineteenth and twentieth centuries and continues to shape it in the twenty-first century. We will explore how religious geographies have shaped and been shaped by broader changes in Ireland's economy and society in terms of both the short-term shocks and the long-term processes.

The Sources: Ireland's Censuses

The census is an excellent source for exploring socioeconomic geographies and how they change over time. As with many other countries, Ireland took censuses for most of the nineteenth and twentieth centuries. Unlike most other countries, however, Ireland has included data on religion as part of its census since 1861, an indicator that religion was, and remains, of more interest in Ireland than elsewhere. The first census of Ireland was taken in 1821, and censuses were repeated decennially until 1911. After Partition, the pattern becomes slightly more complicated. The next census, in 1926, took place on both sides of the border. In 1936 there was a census in the Free State, while in Northern Ireland a comparable, but more limited, census took place in the following year. After World War II the Free State took a further census in 1946, while Northern Ireland held off until 1951, resulting in the largest discontinuity between the two. Fortunately, after this the dates of the two censuses merged again to take place in 1961, 1971, 1981, and 1991. The final census used in this book occurred in 2001 in Northern Ireland and 2002 in the Republic. Thus we have census data for the island covering most of the nineteenth and twentieth centuries available for intervals of every ten to fifteen years and broadly comparable north and south. The bulk of these statistics from 1821 to 1971 have been digitized as the Database of Irish Historical Statistics, allowing a reexamination of the patterns that they contain.[7]

Censuses contain a wealth of spatial information because they are organized using administrative units such as districts and counties. For the nineteenth century the *barony* was the principal administrative unit used in the Irish censuses. There were around 330 baronies (the exact number depends on the date), with an average population of around twenty thousand. From 1901 baronies were replaced by urban and rural districts, otherwise known as county districts. Although there were a similar number of these, their arrangement was significantly different from baronies, as they explicitly separated urban areas from their rural hinterlands. Data for urban and rural districts continued to be available throughout the twentieth century. Although more detailed data are sometimes available at, for example, the parish or townland level, these are rarely consistently available in digital form, so before 1971 the bulk of the sources for this book are used at the barony and urban and rural district levels. From 1971 for Northern Ireland data are available using a unique arrangement of grid squares. These squares have sides that are 1 km long in rural areas and 100 m long in urban centers such as Belfast. This allows us to explore this period in Northern Ireland in much more detail than in the remainder of the country. One problem, however, is that some data are only available for the thirty-two counties of Ireland, robbing them of much of their geographical detail. The counties, in turn, aggregate to four provinces—Ulster, Connaught, Leinster, and Munster—that are too aggregate for statistical reporting but are useful for geographical description. The main administrative geographies are shown in figure 1.1.

Although the details vary slightly, in general, census data on religion consist mainly of the total Catholic, Church of Ireland (part of the Anglican Communion), and Presbyterian populations. Data on minor religions such as other Protestant groups and Jews are sometimes also available; however, the three main groups provide the overwhelming proportion of Ireland's population. In 1911, for example, only 3 percent of the population did not profess to be either Catholic or from one of the two main Protestant denominations. Unfortunately, data on religion are not always available with the level of detail that would be desirable. Digital data from the censuses of 1871 to 1901 only provide statistics at the county level, and from 1971 censuses in the Republic stopped subdividing the Protestant population into Church of Ireland and Presbyterians at local levels, perhaps reflecting the declining demographic importance in the distinction between the Protestant groups in this part of Ireland. It is important to note that the census records the religion that individuals professed to. It is thus a measure of religious identity rather than religious practice.

A final problem is that the census did not start collecting data on religion until 1861, meaning that there is no information on the pre-Famine period. This problem is partly resolved through the use of the 1834 Commission, which examined "the state of religion and other instruction in Ireland." The Commission provides similar data to the later censuses but is only available in digital form using the Church of Ireland's dioceses, of which there were thirty-two but whose arrangement is significantly different from that of the thirty-two counties.

The Methods: GIS and Spatial History

The census and the 1834 Commission thus provide us with spatially detailed pictures of religion and a range of other social indicators in Ireland through the nineteenth and twentieth centuries. Modern technology, in the form of a Geographical Information System (GIS), provides us with new methods of exploring these data. A GIS combines data such as the statistics included in the census with the boundaries for which those data were published. This provides a number of advantages for historical research.[8]

Fig. 1.1. The main administrative geographies of Ireland showing (*a*) baronies, (*b*) rural and urban districts, and (*c*) counties and provinces. Note that before Partition * Laois and ** Offaly were called Queen's County and King's County, respectively.

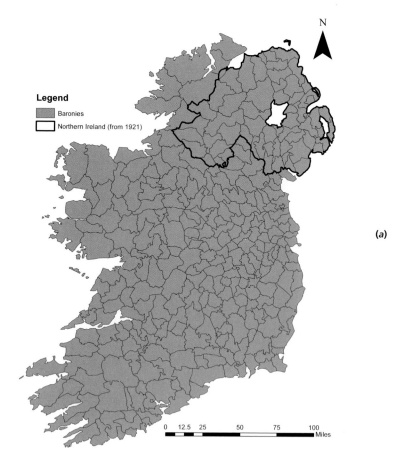

(*a*)

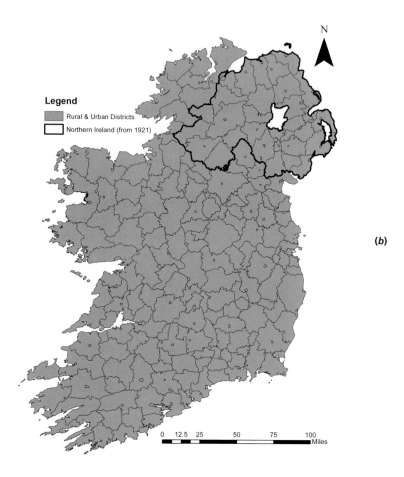

(*b*)

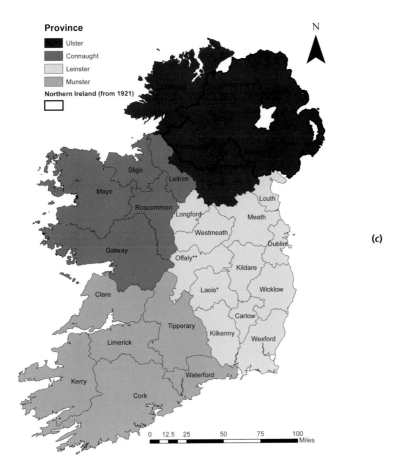

Province
- Ulster
- Connaught
- Leinster
- Munster

Northern Ireland (from 1921)

N

(c)

First, it allows the data to be mapped quickly and easily. Second, it allows data from different sources to be integrated with each other based on *where* they are located. In this book extensive use is made of a technique called *areal interpolation*, which allows the impact of changing boundaries to be removed so that data from several dates can be compared using the same set of administrative boundaries.[9] The third advantage is the ability to make use of analytic techniques that explicitly include location within them.[10]

Beyond these technical advantages, the use of a GIS stresses the importance of space and geography to the historian. This has led to recent calls for the creation of a new field called *spatial history*.[11] The overlap between history and geography in the form of historical geography has a long tradition.[12] What makes spatial history distinct is that it is a form of historical geography that is enabled by, and also limited by, the use of a GIS and other related technologies. These technologies provide tools that allow the researcher to explore space and time in ways that are far more detailed and far more effective in summarizing large amounts of data than has previously been possible. The challenge is to take these tools and use them to create narratives of the different ways in which change occurs in different places at different times.[13] This is important in Ireland because, as we will see, the way that space interacts with religion and other factors is central to an understanding of the ways in which the island has developed. Spatial patterns and how they change are complex, so more traditional historical geographies have been limited in their ability to make full use of the spatial and temporal detail available to them. Although a GIS does much to help us overcome this limitation, the sources that we use are numeric and aggregate; thus, while they are excellent for describing patterns and how they change, these sources are more limited in their ability to provide separations.

A Spatial History of Ireland

The book primarily covers the period from 1821 to 2001/2002, the first and last censuses for which data are currently available for the whole island, although chapter 2 sets the scene by discussing the much earlier geographies of the plantations. Despite the dates being defined by data availability, there are also strong academic justifications for using them. The period from 1821 to 1841 relates to the pre-Famine period, when, superficially at least, population trends in Ireland were similar to those found elsewhere in what was then the United Kingdom. 2001/2002 provides what may well turn out to be the end of a major chapter of Irish religious history on both sides of the border. In the Republic the labor demands of the Celtic Tiger meant that a country famous for its emigration was about to open its borders to an influx of emigrants from eastern Europe and elsewhere who did not conform to the ethnoreligious identities of the indigenous population. The 2000s would also see the Catholic Church embroiled in scandals that may further weaken its hold in Ireland and increase the secularization that is already happening. In Northern Ireland the Belfast or Good Friday Agreement of 1998 marked the end of thirty years of conflict, despite the fact that the negotiations and—to a far lesser extent—acts of violence have continued. This has meant that Northern Ireland's economy has had the opportunity to begin to revitalize; however, the two communities remain in many ways as divided as ever. Again, secularization and immigration are beginning to have impacts on the religious divide, impacts that were not present through the previous two centuries.

In writing this history we have tried to bridge the gap between the traditional historical narrative, stressing the chronology of events, and an atlas format. Atlases are good at presenting geographical patterns in an attractive way but tend to do this at the expense of a coherent narrative.[14] This book's narrative is concerned with the spatial patterns of religion and society and how these have changed, but it is told using a larger number of maps and other diagrams than would be expected in a historical monograph. In this way it stresses the importance of geographical as well as temporal change.

While a GIS is well suited to producing the maps that form the backbone of this approach, paper is not good at publishing it. Page size, a lack of full color, and a lack of interactivity all limit what can be achieved in a traditional book format. A book, however, still has certain advantages, not least that many people would still prefer to read a page than a screen. For this reason we have compromised between book and electronic formats. This book contains a full narrative account illustrated with many maps in color. The electronic version provides the same maps plus additional resources.

Most of the maps we use include the border between Northern Ireland and the Republic/Free State. This is partly for orientation purposes but also because one of the key themes of this book is how Ireland evolved to the position it is in today. The border was defined by the religious, social, and economic geographies that preceded it and has helped to redefine and reinforce these geographies ever since. It thus seems sensible to include it on maps that predate its existence in part to show that even if northern parts of the island appear different from the rest, there is rarely a simple relationship between these patterns and modern political divisions.

The chapter structure of the book and of the electronic supplement is identical. Chapter 2 describes the plantation geographies; chapter 3 talks about the pre-Famine period, establishing the relationship between the religious geographies of this period and the plantation geographies and exploring the broader trends in Irish society at the time with an emphasis on population growth and agriculture. Chapter 4 explores in detail the impact of the Famine, particularly the period from 1841 to 1861. Chapter 5 then examines the trends—economic, political, and religious—that increasingly divided the island between 1861 and 1911. Chapter 6 looks in detail at the period between 1910 and 1926, exploring how a movement for Home Rule led so suddenly to the partition of the island into two parts along explicitly religious lines that were also related to the political and economic divides on the island. Chapter 7 continues to explore the island as a whole, looking at how it became increasingly divided through much of the twentieth century between an economically successful Northern Ireland and the Free State (later the Republic), which was stagnating economically and demographically. The last few decades have seen a complete change in fortunes for both parts of the island. The Republic became one of Europe's most dynamic economies as the Celtic Tiger, as described in chapter 8. North of the border, however, saw dark times, as the manufacturing industries on which the economy depended went into decline, and the Troubles led to over 3,500 deaths between July 1969 and December 2001 in politically motivated violence that was often explicitly sectarian in character.[15] Three chapters are devoted to this, one looking at socioeconomic change in Northern Ireland, one looking at violence in Northern Ireland during the Troubles, and the final one focusing on Belfast as a microcosm of both the Troubles and the wider processes in action at the time.

What becomes apparent is that an understanding of how Ireland developed over the past two centuries requires an understanding of the major themes that have shaped it, economy, society, politics, and religion being prevalent among them. Time is also clearly important, as recognized by the chronological chapter structure. Time brings with it long-term trends—industrialization and urbanization, deindustrialization and counterurbanization, and short-term shocks such as the Famine, Partition, and the onset of the Troubles. Time also brings the past—history with its lessons, identities, resentments, and prejudices—and the future—its plans that succeed and fail to greater and lesser extents. Geography, however, is of critical importance, as space is where all of these forces occur, develop, and interact. There is not just one story of Irish religion and society over the past two centuries, there are many interweaving stories of how the many aspects of religion and society have interacted together in different ways, at different times, in different places. The story of the north is different from the story of the south, and the story of the east is different from the story of the west. The major cities, Dublin and Belfast, are different from each other and from the towns and the countryside. These places themselves—north, south, east, west, city, town, and countryside—are not static but constantly evolving, both shaping and shaped by the themes of society and religion.

This book therefore tells the major narratives of religion and society and how they have changed in Ireland and changed Ireland over the last two hundred years. The aim in writing it has been to identify the main stories that the major different places have to tell. Given the limitations of a text, however, it is impossible to tell all the stories of all the places. We hope that the accompanying maps and electronic resources allow readers to explore these stories in more detail for themselves.

The Plantations: Sowing the Seeds of Ireland's Religious Geographies

2

The major plantations of Ireland, which were put in place during the sixteenth and seventeenth centuries, were an attempt, or a series of attempts, to establish a Protestant population from England and Scotland in Ireland. This occurred for both political reasons—Protestant England was worried about the threat that Catholic France and Spain could pose through Ireland—and economic ones, in particular due to the close trading ties between southwestern Scotland and northeastern Ireland. The plantation period is outside the temporal range of this book, and the sources on which much of the remainder of the book is based do not exist for this time. We have, however, included a brief description of the events that occurred and the geographies that they established, since, as chapter 3 will describe, their legacies lasted until the early nineteenth century and therefore provide the foundations of much of what was to follow. Indeed, the events of this period left spatioreligious patterns that continue to have an influence to this day.

Figure 2.1 summarizes, in general terms, the geographies of the major plantations of Ireland. Major plantations were established in the Pale—the area around Dublin—and in Ulster. There were more sporadic attempts to create plantations in Munster, and no attempt was made to plant western parts of Ireland. As we will see, early nineteenth-century geographies still showed concentrations of Protestants in Ulster, the Dublin area, and west Cork, along with very low concentrations of Protestants in the west. Many of these patterns persisted to the end of the twentieth century. There are also some clear differences, as some of the areas that were planted, particularly in parts of Munster and some of the later Jacobean schemes, now have very small Protestant populations. Explaining these continuities and changes requires an understanding of the geographical and social impacts of the plantations, as they are, in many ways, the origins of the divisions and interdependencies that will be discussed in much of the rest of the book.

Phrases like "the plantation of Ireland" and depictions like figure 2.1 are in many ways misleading. They imply that the process of colonization was an organized, long-term plan with clear, overarching objectives that resulted in neatly contained spatial areas. The plantation system was anything but neat—it was more often reactive than proactive, it was disorganized and incoherent in its approach, and the geographies that it left were often disjointed and contested. Changing priorities, circumstances, and values over time mean that it is impossible to view the process of colonizing Ireland as methodologically and ideologically consistent.[1] This inconsistency of approach and impact goes a long way to explaining Ireland's contemporary religious geographies and has resonated down through the centuries.

Fig. 2.1. Major plantation schemes and areas of English/British influence in Ireland in the late sixteenth and early seventeenth centuries.

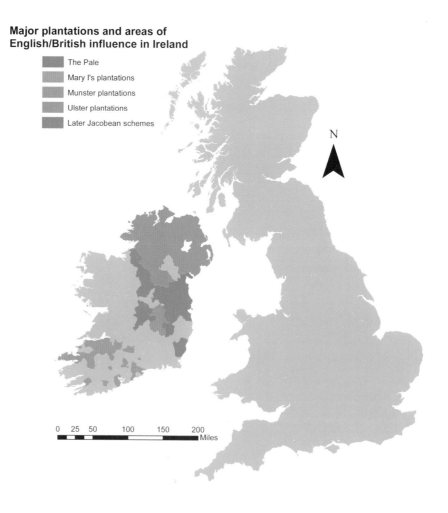

Major plantations and areas of English/British influence in Ireland

- The Pale
- Mary I's plantations
- Munster plantations
- Ulster plantations
- Later Jacobean schemes

N

0 25 50 100 150 200
Miles

Background

The late medieval/early modern period marked a fundamental turning point in Anglo-Irish relations. Prior to the reign of Henry VIII, English control of Ireland had been a largely nominal affair. Power had been exercised through Crown representatives who were members of the aristocratic elite known as the "Old English." The Old English were the descendants of the Norman families who had moved into Ireland in the late twelfth century, the first serious attempt at Ireland's colonization. Over the centuries, successsive English monarchs had become increasingly suspicious of the motives and loyalties of Old English families, such as Munster's House of Desmond, as they became not only more powerful but also progressively more gaelicized in their attitudes and behavior, intermarrying with influential native Irish families.[2]

Concerns about the adoption of the Irish language and modes of dress may have been superficial, but when the Earl of Desmond became directly involved in the internal vicissitudes of English politics by supporting the ill-fated House of York during the Wars of the Roses, his action indicated to the reigning Tudor monarch, Henry VII, that Ireland presented a significant threat to English security.[3] His son, Henry VIII, regarded lordship of Ireland as part of his birthright and believed that if he did not exercise his right of inheritance, it was possible that one of the more powerful earls would try to wrest it from him.[4] This is precisely what happened when

"Silken Thomas," the son of the ninth Earl of Kildare, rebelled in 1534, an act that led Henry VIII to declare himself king of Ireland eight years later and signaled the beginning of a much more aggressive English colonial posture in Ireland.[5]

The subsequent process of plantation, which started in the mid-sixteenth century, must be understood not only in the context of the potential threat of instability that successive monarchs regarded Ireland to represent but also in the light of the theological revolution that was to redefine Europe's religious landscape and to more profoundly radicalize relationships between the two islands. The fact that Ireland remained Catholic while Britain turned toward Protestantism introduced a new and complex dimension into Anglo-Irish relations that would increase levels of mutual distrust and suspicion. Throughout the reigns of Elizabeth I and James I, Catholicism and its followers were increasingly viewed as treacherous and subversive at least in part because of the wider European conflicts with France and Spain. It was quite apparent by the late sixteenth century that the Reformation would bypass Ireland so if Ireland would not become Protestant, then Protestants would have to be brought to Ireland.[6] The only problem was that importing Protestants into Ireland forced the natives to confront the everyday realities of the Reformation and a new religious dispensation under which they had effectively been cut adrift.

The Midlands Plantation of Mary I (1556)

Figure 2.2 summarizes the waves of plantations in Ireland. The first major plantation scheme was initiated by Mary I. It occurred in the midlands from 1556 in the area that corresponds to modern-day Counties Laois and Offaly, known as Queen's County and King's County respectively, before Partition. In the mid-sixteenth century Dublin and the areas surrounding it, collectively known as "the Pale," constituted the main area of English authority in Ireland. During the 1550s the Pale had come under increased attack from native chiefs, and Mary I, concerned by the threat to the center of Crown influence in Ireland, sought to secure the borders of the Pale by the establishment of two largely military settlements, centered on the towns of Maryborough (modern-day Port Laois in Laois) and Philipstown (now Daingean in Offaly). The scheme was only a partial success. It largely halted rebel attacks on the Pale and also achieved a secondary objective of improving communications across the boggy central plains of Ireland. However, it did not diminish rebel attacks but rather acted to deflect them away from the Pale and toward the two new garrison settlements. The ongoing danger meant that the colony had difficulty in building a civilian population and remained a quasi-martial settlement. Rather than becoming self-financing, as had originally been envisaged, these places thus continued to be a burden on the English Exchequer.[7]

The midlands plantation was the first organized attempt to settle a foreign population in Ireland and met with limited success. The accession of Mary I to the throne and the temporary restitution of the Catholic faith to its former status in England altered the impetus of the scheme in that it went from being an exercise in altering religious demographics— as probably intended by the radical though short-lived Edward VI—to a

Fig. 2.2. Tudor plantations of Ireland.

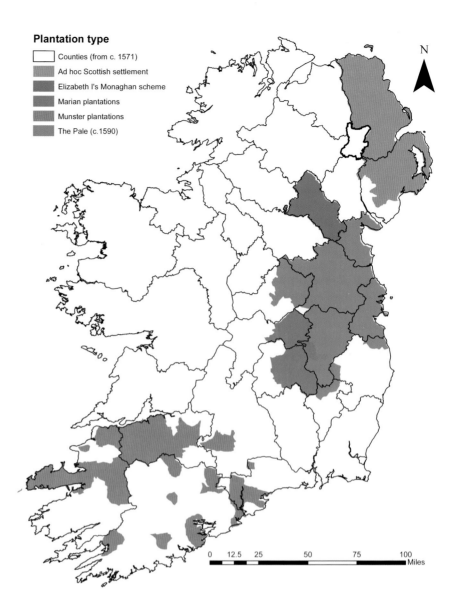

Plantation type

- Counties (from c. 1571)
- Ad hoc Scottish settlement
- Elizabeth I's Monaghan scheme
- Marian plantations
- Munster plantations
- The Pale (c.1590)

N

0   12.5   25   50   75   100
Miles

straightforward buttressing of the Pale.[8] The progress of the Reformation in England at this time complicates the picture. The fact that Mary I was a Catholic who intended to restore the Catholic faith should not be overemphasized. The reality was that it was already too late to undo most of the reforms to the church that had taken place under Henry VIII and Edward VI and that were largely Protestant in character.[9] Under Mary we thus see that ethnic or national identity was closely related to religious identity; thus, land confiscated from the natives was to be settled by "Englishmen born in England or Ireland."[10] Whether the test was one of ethnicity or religious belief would become increasingly irrelevant: what mattered in Ireland was that religion and ethnonationality were beginning to separate into two mutually exclusive and antagonistic identities.

The Munster Plantation (1585)

Retribution and reaction provided the motivation for the next phase of plantation. After the 1585 defeat of a rebellion by the Earl of Desmond, an Old English Catholic, his extensive lands across the province of Munster were confiscated. Spoliation had decimated the population of Munster and brought a terrible famine upon the survivors.[11] Settlers, largely from the west of England, were brought in to repopulate the region under the stewardship of people such as Edmund Spenser and Richard Boyle, the first Earl of Cork.

The piecemeal nature of the plantation in Munster was reflective of the pattern of land forfeited from the Earl of Desmond and is evident by referring back to figure 2.2. The project's implementation was as unsatisfactory as its geography. It was a painfully slow process. By 1588 only three thousand people had been settled on the disparate former Desmond estates; ten years later this number may have reached twelve thousand. The delays encouraged former residents to restate their claims to the land through legal means or others simply to squat on vacant plots.[12] In 1598 the plantations suffered fresh devastation as the Desmond heirs, inspired by the victory of Hugh O'Neill over the English at the Battle of the Yellow Ford, rebelled in an attempt to regain their former lands. Spenser himself was lucky to escape with his life after his estate at Kilcolman in west Cork was attacked and burned.[13] O'Neill was the most powerful of the Gaelic leaders, and his defeat and subsequent departure created a power vacuum in Ulster that in turn paved the way for the next major phase of colonization.

The Munster plantation recovered, and by 1641 the settler population was estimated at twenty-two thousand, growing from around five thousand in 1611.[14] The Munster plantation has left a permanent mark on Ireland's religious geography, with a significant Protestant minority being enumerated in the south and west of County Cork into the twentieth century, but its long-term impact in the rest of Munster is much harder to discern. One of the most striking features of the Munster plantation was the degree of mixing between native Catholics and Protestant newcomers. Intermarriage was not only decriminalized but later actively encouraged as a means of drawing the native Irish more deeply into the wider English, or by then British, social and cultural world, thereby advancing the "civilizing" mission of colonization. This policy was so successful that it became impossible to record all instances of mixed marriage.[15] It appears, therefore, that in this part of Ireland relatively harmonious conditions prevailed despite the religious differences. This would stand in stark contrast to the situation that would develop in Ulster.[16]

The Ulster Plantation (1603)

The colonization that has had the most profound and enduring impact upon Ireland's history occurred in Ulster and started during the reign of James I at the beginning of the seventeenth century. The Ulster plantation was the most ambitious in scale and in its ideological underpinnings. Ulster had long been considered the most volatile and underdeveloped province of Ireland, with a tradition of powerful native clans, most notably, the aforementioned O'Neills.[17] Ulster's position meant that strong ties of kinship with the Catholic Highland clans of Scotland existed across the North Channel, presenting an enduring strategic threat to English authority in the dominion. The Ulster plantation followed the Nine Years War of 1594–1603, which was particularly concentrated in the province and led to the defeat of the old Gaelic order. It also showed the threat that Ulster presented to English interests and offered an opportunity to ensure that it would never provide the seedbed for revolt again.

The main scheme of the Ulster plantation was focused on breaking up the old centers of Gaelic influence in the west of the province and thus applied to all counties west of Lough Neagh and the River Bann. Most of the land was portioned out amongst English and Scottish "undertakers," entrepreneurs who undertook to settle their allotments with Protestants. The rest of the land would go to army veterans known as "servitors," native Catholics who had shown their loyalty to the Crown during the Nine Years War, and in the form of endowments to the Church of Ireland and the new University of Dublin, to become better known as Trinity College.[18] The plan for the county of Coleraine was different. Due to a lack of money, James I tried to hand responsibility for its settlement to the powerful guilds of the City of London, and it effectively became a private enterprise henceforth known as Londonderry.[19] Counties Antrim and Down were excluded from the formal colonization process, as they had already been the subject of highly successful private plantation schemes under Hugh Montgomery and James Hamilton.[20]

Figure 2.3, although seemingly detailed, actually presents a very generalized plan over most of the six escheated counties that formed the organized plantation.[21] None of these areas was completely populated by just one of the settler groups; each area contained all groups in varying proportions. More importantly, the new plantations also contained substantially more Catholics than was originally intended. The initial aim had been to keep the natives on the open ground within sight of the servitors, who were introduced not simply as a reward for military service but also as a means of keeping an eye on the rebellious Catholic population.[22]

While the natives could hold land, the patents that they received were often impermanent and subject to arbitrary revision or withdrawal.[23] The corollary of this was that the position of Catholic tenants on native lands was equally imperiled. Although Catholics could rent land from Irish grantees or servitors, they paid higher rents for not taking the Oath of Supremacy, which ordered allegiance to the British monarch as head of the Churches of England and Ireland. Where land was awarded to native grantees, it was rarely the land that their clan had historically occupied.[24] This goes to the ideological heart of the project—it did not simply try to supplant groups of differing religions and allegiances but actually attempted to destroy the relationship between space and ethnicity that the old Gaelic territories represented. This, combined with the insecurities of tenure for native grantee and tenant alike, led to "a dislocation that could be not only geographical but psychological."[25] In turn, this led to the nurturing of animosities and inequalities at a local level that exploded into violence in the rebellion of 1641. Even in Monaghan, which had been granted in whole to native tribes in 1591, dislocation was apparent—by 1640 those same clans held less than half the territory in the county.[26]

This territorial usurpation of the Gaels was achieved not by overt aggression but by the imposition of an alien system of tenure and free-market economics in which the odds were ethnically stacked against them.[27] Meticulous rules existed as to how and to whom land should be parceled up;

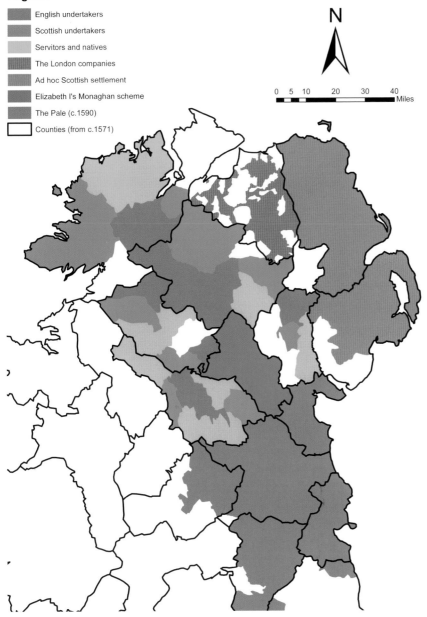

Fig. 2.3. The Ulster plantation, 1609–1613.

Legend

- English undertakers
- Scottish undertakers
- Servitors and natives
- The London companies
- Ad hoc Scottish settlement
- Elizabeth I's Monaghan scheme
- The Pale (c.1590)
- Counties (from c.1571)

N

0 5 10   20   30   40
Miles

however, even from the earliest stages it was quite apparent that the acreages on the patents bore little relation to those on the ground, as the surveys woefully underestimated the size of the estates that had been created. One might initially think that this was serendipitous, given the grievances that would be fostered by attempting to settle a greater population within a finite area of land, but in fact it created fresh problems that threatened to compromise the integrity of the entire project. This is ironic, as an extremely detailed spatial hierarchy existed within the Gaelic order in which townlands, tates, and ballybetaghs formed an efficient geographical framework by which the plantation could have been organized.[28] The inaccuracy of the surveys meant that the undertakers, who numbered only about one hundred and whose role was absolutely central to the plantation enterprise, were left in control of substantially larger areas than they had envisaged. This meant that the obligations placed upon them to develop their estates under the terms of their patents were also extended. As a result, these undertakers began allowing Catholic tenants to settle on their land. This was initially a short-term arrangement; however, it soon became apparent that the natives were a more abundant and economical source of income than settlers from Britain. Lacking either the inclination or the skills, the English tended not to plant their land, preferring instead to let it to the Irish, who in turn were reluctant to make the kind of temporal investment in the soil that arable farming required because they had no idea if or when eviction could occur.[29]

A symbiosis of mutual dependence had thus developed between the English and the Irish. The English were dependent on the Irish for rents, and the Irish were dependent on the English for land, yet the insecurity of the arrangement for both parties was deleterious to the economic, infrastructural, and moral objectives of the plantation.[30] This arrangement continued despite the financial penalties imposed by London which, in frustration and dogged expectation, kept extending the deadline by which Catholics had to be evicted from those estates, first to 1611, then to 1618, and finally coming to a compromise a decade later that resulted in Catholics staying put, but again at a price. The government finally concluded that full racial separation would probably precipitate the plantation's collapse. Crucially, this meant that the British government had capitulated on the ideal of absolute religious segregation.[31]

The emphasis laid down in the original plantation plans for religious segregation, albeit never fully realized, marked the start of a polarization that has lasted down through the centuries and become one of the defining characteristics of Ulster society. However, even in Ulster instances can be found of the partial and variegated nature of the plantation experience. In one example, the Earl of Abercorn, a Scotsman and favorite of King James I, was given a large grant of land outside Strabane in west Tyrone, yet he was a Catholic, and in 1649 his son was excommunicated from the Presbyterian Church of Scotland for his family's continued recusancy.[32] In another, the Archdales of Norfolk left England for County Fermanagh accompanied by the Eves family, who were Catholic and who prospered in Ulster under the patronage of their employers. In both cases, we have the bizarre situation of Irish Catholics being supplanted by their British co-religionists for the purpose of an ostensibly reformist plantation.[33] Examples such as these highlight the fact that religion and ethnic identity were not always aligned, and, in these particular cases, ethnicity seemed to override religion.

For the Ulster plantation to have fully succeeded it would have to have imposed a completely new religious geography that would have permanently removed the threat of rebellion by assimilating the native population into a new social, economic, and religious landscape. This ambitious objective was only partially achieved: new religious geographies were successfully imposed on Ulster and remain to this day. As chapter 3 will describe, much of the area across west Ulster still formed the heartland of the Church of Ireland in the modern era, while the northeast counties of Antrim and Down still bore the imprint of organized and ad hoc migration of Scottish Lowlanders who brought them the Presbyterian form of Protestantism in the late sixteenth and early seventeenth centuries. The economic and political dispossession of the Catholic population, which was mixed and displaced between the Church of Ireland and the Presbyterian Church, in Ulster started a festering resentment that endured for the next three and a half centuries.[34]

The creation of the new religious geographies of Ulster brought Catholics and Protestants into closer physical proximity, yet on the social, economic, and religious planes the scheme contrived to create almost parallel societies. The position of Catholics would continue to remain largely subordinate, and the religious geographies would remain both stable and contested as Catholics and Protestants developed an acute sense of religious territoriality over the long term.[35] For these reasons, the Ulster plantation would fail in its philosophical ambition to Anglicize the Catholic population of Ulster in anything other than language and would provide the genesis for generations of spatioreligious conflict.

The Cromwellian Land Settlement (1652)

The partial success of the Ulster plantation in imposing new religious geographies came at a high price in both human costs and long-term consequences. In 1641 the Catholics of Ulster rebelled and killed several thousand of their Protestant neighbors. This event, along with the 1689 Siege of Derry, would calcify into Protestant mythology the image of Catholic savagery and barbarity.[36] In one of the great ironies of Irish history, Oliver Cromwell took his revenge for the rebellion not in Ulster but in towns in the south which had not participated in the revolt, with massacres in Drogheda and Wexford in 1649.[37] The grievances that resulted were nurtured throughout Catholic Ireland, and nowhere more so than in Ulster. In terms of Ireland's religious geographies, the subsequent Cromwellian land settlement of the 1650s also provided the final missing piece of the complex jigsaw that has done much to define the differing spatial distributions of the three main religious groups.

Despite the diminution of the status of Catholics, which had been an ongoing process since the start of the plantations, in 1641 they still owned 59 percent of Ireland's land.[38] It had long been recognized that landownership was the basis of power in Ireland, and, through the largest redistribution yet envisaged, Cromwell hoped that the threat posed by Ireland's Catholic population could finally be neutralized.[39] Estimates about the numbers of deaths during the rebellion vary, although it has been estimated that between two thousand and four thousand settlers died. Subsequent retaliatory attacks against the Catholic population brought the overall death toll higher still and created a new low of degeneracy in religious conflict in Ireland, exemplified in the grim aphorism that "nits make lice" being given as a not uncommon justification for the killing of Catholic children in Ulster.[40]

The death toll is perhaps less significant than the impact of the rebellion in acting as a justification for Cromwell's military campaign in 1649–50 and the land settlement that followed.[41] Cromwell held all Catholics responsible, but his response reflected ambivalence toward the "deluded and seduced" peasantry, on the one hand, and the landed native Irish who had actually caused the rebellion, on the other.[42] There was truth in this analysis, as, curiously, it was those Irish who had retained land in Ulster because of their loyalty to the Crown during the Nine Years War who actually chose to revolt in 1641.[43] Irish lords like Owen Roe O'Neill (a nephew of the great Hugh), Maguire, and O'Reilly actually owed their position to the plantation of Ulster. They rebelled in the name of the king because England was on the brink of the Civil War, and they feared that a parliamentary coup in London would cost them their privileged position.[44] The fact that these natives were ardent royalists would not have endeared them any further to Cromwell.

In the end they did lose their privileged position as Catholics were forbidden from owning any land east of the River Shannon, a physical barrier between the west of Ireland—mainly Connaught—and the rest of the island. So it was the native owners of land, rather than the great bulk of the Irish, who were targeted by this last great attempt to redefine Ireland's religious landscape. The impact of the land settlement is shown in figure 2.4.[45] Cromwell intended to clear all Catholics from east of a line stretching from the River Barrow to the River Boyne, an area that includes the Pale and much of the rest of eastern Ireland south of Ulster. A lack of British settlers meant that such a plan was never practical, and the new Protestant landowners quickly argued for the retention of a Catholic working class.[46] Just

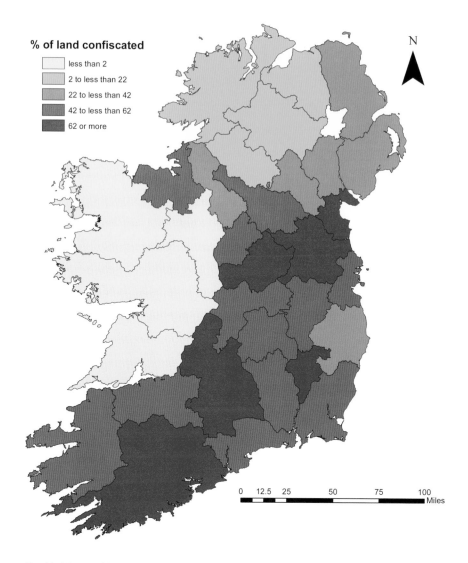

Fig. 2.4. The Cromwellian land settlement.

as with the Ulster plantation, while the idea of religious segregation was at the core of the theory, in practice such an arrangement was unworkable, and whatever animosities existed, a degree of economic interdependence had continued to develop between Catholics and Protestants. For landed Catholics everything changed as their fertile holdings in south Munster and elsewhere were forcibly exchanged for rocky, windswept acres in Connacht or Clare.[47] The poor were perhaps less altered once the initial expulsions had proved both fruitless and pointless.

Conclusions

It is impossible to overestimate the impact of the plantations in defining Ireland and its religious geographies in the modern period. The plantation of Ulster in particular succeeded in transforming the religious complexion of the island. The Cromwellian land settlement permanently marginalized Catholics economically and theologically while cementing the Protestant ascendancy.[48] Religious segregation was often at the heart of the colonization ideal and was closely linked to ethnic identity, political allegiance, and socioeconomic status. This resulted in geographical as well as social segregation. However, this period also showed that the two communities were mutually dependent as attempts to separate the two geographically were undermined by Protestants themselves due to their economic reliance on Catholic labor. This was thus the real start of what the historian F. S. L. Lyons has described as an anarchy of the heart and mind, "of the collision . . . of seemingly irreconcilable cultures, unable to live together or to live apart, caught inextricably in the web of their tragic history."

> Out of Ireland we have come;
> Great hatred, little room,
> Maimed us at the start.[49]

3 Religion and Society in Pre-Famine Ireland

The first population censuses were taken in Ireland in 1821, 1831, and 1841, but while they contain geographically detailed information about the distribution of the population, they did not include any information on religion. The Commission of Public Instruction, Ireland, taken in 1834, does, however provide us with data on religion for this period. The Commission was instigated by the nonconformist Whig government in London, which sought to use its results to assail the privileged position of the established Church of Ireland.[1] Prior to this survey the extent of the Catholic majority in Ireland had been grossly underestimated, and the desire to uncover the demographic strengths of Ireland's religions was fueled by a strong desire among Protestant evangelicals to proselytize the majority group.[2] This chapter uses the Commission and the early censuses to explore the geographies of religion and society immediately before the Great Famine of the late 1840s. They show that Ireland had both similarities with and differences from the rest of Europe. As with other European countries, the population was starting to grow rapidly; however, in Ireland a lack of industrialization meant that rural population pressures were growing. The island also already had clear and polarized spatioreligious patterns that still closely followed those laid down during the plantations. Presbyterians, primarily the descendants of private Scottish planters and ad hoc migrants, were concentrated in the northeast of the island. The Church of Ireland had a much more fragmented pattern, being spread along south Ulster and east Leinster and reflecting the relative lack of success of the plantations in many of these areas. The rest of the island was overwhelmingly Catholic.

The first part of this chapter uses the 1834 Commission to describe the geographies of religion of the early nineteenth century and explores the extent to which these geographies had their roots in the plantation period. The second part is more forward-looking, exploring how nineteenth-century trends such as population growth and industrialization were emerging in this period and how Ireland was in some ways similar to other parts of Europe and in other ways very different from them. The differences, in particular, were to have deadly consequences when famine struck and also ensured that the Great Famine would have impacts in Catholic Ireland different from those in more Protestant parts.

Religion in 1834

The 1834 Commission provides us with the first detailed survey of the distribution of Irish religion. The survey, available in digital form for the Church of Ireland's dioceses, of which there were only thirty-two, does not provide the level of spatial detail that is available in later sources. Nevertheless, this source does present the best available snapshot of religion in the pre-Famine period. It shows that across the island 80.9 percent of the population professed to be Catholic, while the Church of Ireland was slightly the larger of the two major Protestant groups, with 10.7 percent to the Presbyterians' 8.1 percent. "Other Protestants" made up only 0.27 percent of the population. While these data must be treated with a little caution, they have been described as being "remarkably accurate for an early nineteenth-century statistical study."[3]

The three maps in figure 3.1 show the spatial distribution of the three largest religions as recorded in 1834. Each of the three groups has a distinctive geography. Catholics were by far the largest individual religion in Ireland, predominating across the three southern provinces of Connacht, Munster, and Leinster. Over the bulk of this area Catholics accounted for over 90 percent of the diocesan population. Even in the extreme northeast of the island, where they were least represented, they still accounted for over 25 percent of the population of the dioceses of Connor and Down, which approximate to the counties of Antrim and Down, respectively. The Church of Ireland has a less concentrated pattern, with no diocese having over 40 percent of its total population being drawn from this religion, yet a clear geography is evident, straddling south Ulster and north and east Leinster. For the Presbyterian Church the distribution is again markedly different, being heavily concentrated in the extreme northeast of the island. While their proportion of the population of a diocese never exceeded 60 percent, Presbyterians did constitute over 40 percent in all of the dioceses facing Scotland and over 50 percent of the population of Connor and Down.

Religious Change since the Plantations

Chapter 2 described the longer-term historical background to Ireland's pattern of religious settlement. The enumeration of 1834 provides the first real opportunity to gauge how those religious geographies had either altered or remained the same over the intervening two centuries, albeit at a highly aggregate scale. The 1834 survey provides a useful picture of where the patterns of religious settlement laid down by the plantations persisted and where they had waned. Figure 3.2 compares baronies that had undergone some sort of organized or large-scale colonization during the sixteenth and seventeenth centuries with the diocesan religious geographies of 1834. The differences in scale between over three hundred baronies and the thirty-two dioceses mean that the patterns must be treated with caution, as all we are able to conclude is that a planted barony lay in a diocese that had a particular proportion of Protestants in 1834. We are not able to determine what the proportion of Protestants within that barony was. Despite this, the map gives us an insight into the varying spatial patterns of persistence and decline in the religious geographies that the plantations had established in Ireland. The highest levels of Protestant persistence existed in Ulster, specifically northeast Ulster. Other areas of persistence occurred in east Leinster and west Cork. In other parts of Leinster and Munster, the Protestant population of planted areas appears to have largely disappeared.

Fig. 3.1. Religion in Ireland in 1834 at diocese level: (a) Catholics, (b) Church of Ireland, (c) Presbyterians.

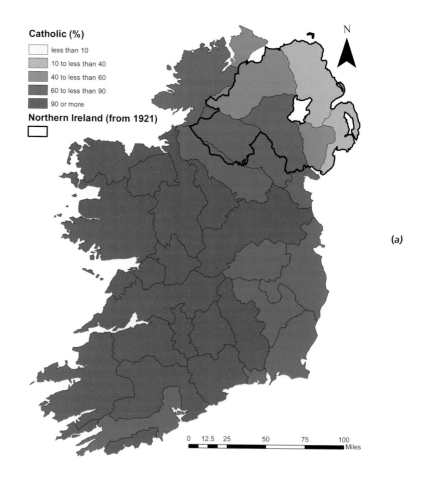

Catholic (%)

- less than 10
- 10 to less than 40
- 40 to less than 60
- 60 to less than 90
- 90 or more

Northern Ireland (from 1921)

0   12.5   25   50   75   100
Miles

(a)

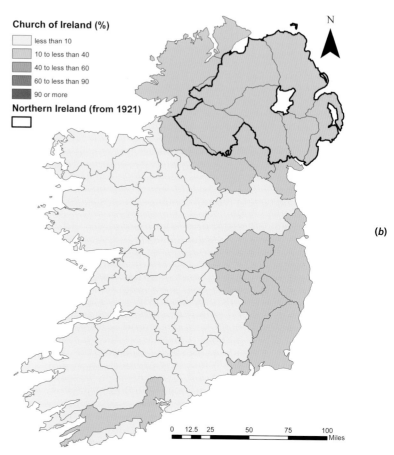

Church of Ireland (%)

- less than 10
- 10 to less than 40
- 40 to less than 60
- 60 to less than 90
- 90 or more

Northern Ireland (from 1921)

0   12.5   25   50   75   100
Miles

(b)

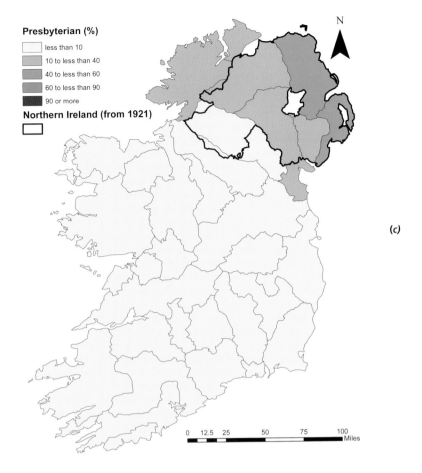

Presbyterian (%)

less than 10
10 to less than 40
40 to less than 60
60 to less than 90
90 or more

Northern Ireland (from 1921)

N

(c)

0 12.5 25 50 75 100 Miles

This pattern reflects a number of important points. First, it is evidence of the ambitiousness and extensiveness of the Ulster scheme in comparison to earlier attempts to colonize rebellious parts of the south of the island. In Leinster and Munster, the native clans were not displaced to the same extent; thus, the populations were always more mixed, and this was still reflected in the 1834 geography.

The second point is a development of this. It is clear from the map that either much of the Munster plantation had withered or the initial influx of Protestant settlers had never been of the same demographic concentration as was achieved with the plantation of Ulster. Figure 3.2 illustrates that by 1834 much of the plantation had failed to take root outside of a limited area of south Cork and therefore had not altered significantly the long-term religious geographies of the southwest. By the eve of the 1641 Rebellion, the entire settler population in Munster was estimated at only 22,000, fully sixty years after the colony's establishment.[4] Aidan Clarke estimates that the adult British-born (not simply settler) population in the six escheated counties of Ulster was already about 13,000 in the early 1620s, while the same group in the privately planted counties of Antrim and Down was somewhere around 7,500, meaning that Ulster had reached the same level of colonial development within approximately one-third of the time.[5] Furthermore, the highly uneven nature of Ireland's population distribution in the early seventeenth century meant that it was easier to affect Ulster's religious demography than Munster's. In 1732–33 McCracken has estimated that Munster was Ireland's most populous province, with 31 percent of an approximate total population of 2.8 million for the island, while Ulster was the least populated

Fig. 3.2. Comparing Protestantism in 1834 with the plantation. The map compares 1834 diocesan religious populations with a barony-level geography of the plantation. Areas shaded in the darkest shade are planted baronies that by 1834 lay within a diocese that was over 50 percent Protestant. Unplanted baronies are unshaded. The difference in scale between baronies and diocese means that these results must be treated with caution.

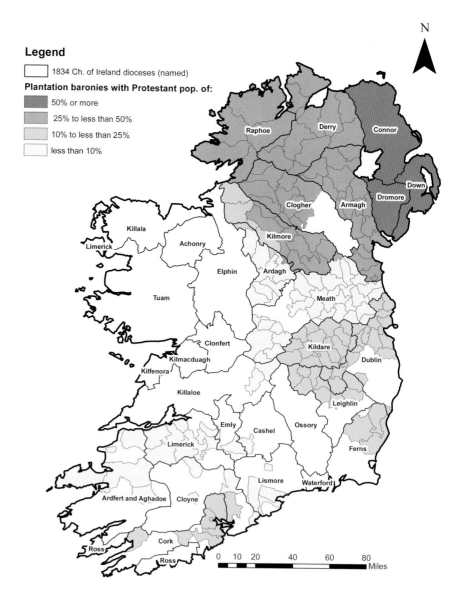

Legend

☐ 1834 Ch. of Ireland dioceses (named)

Plantation baronies with Protestant pop. of:

■ 50% or more

■ 25% to less than 50%

▨ 10% to less than 25%

☐ less than 10%

province, with only 12 percent of the total.[6] While these figures should be treated with caution, it is widely accepted that Ulster was Ireland's least populated and least developed province at the start of the period, and it is probable that, if anything, its share of Ireland's population may have been lower than 12 percent in the 1620s, given the province's recent experience of war and despoliation.[7]

The early failure of the Munster plantation is also apparent in the locations of Protestants who contributed to the book of evidence compiled against the native Catholic Irish who rebelled in 1641. Figure 3.3 contrasts the locations of the original Elizabethan plantations with the locations and numbers of "depositions," as they became known, and the concentration of Protestants in 1834 at the diocese level. The depositions show that the Protestant population of Munster had already become heavily concentrated in relatively small areas of Munster by 1641,[8] contrasting with the much more dispersed pattern of the original Elizabethan colonies. So it would appear that the pattern of Protestant settlement in Munster that could be observed in 1834 was one that had probably taken shape fairly shortly after the initial late sixteenth- and early seventeenth-century schemes. It is also true

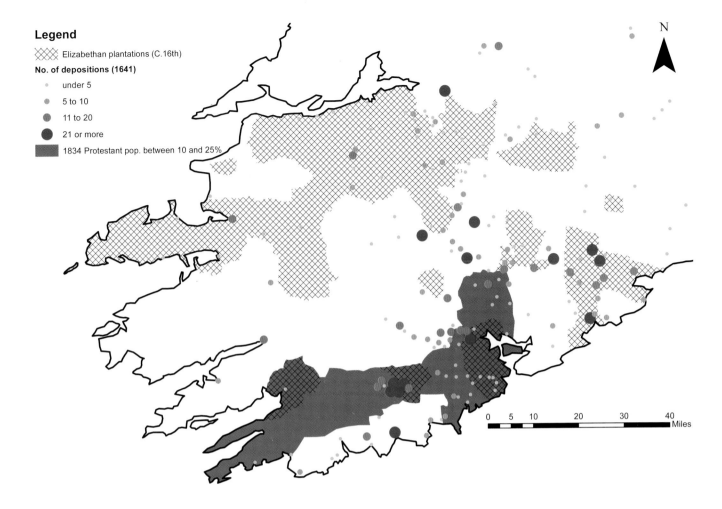

that the settler populations in Ulster did not necessarily stay in the same places to which they were initially assigned. While the principal motivation may have been pecuniary in order to take advantage of the most profitable land, another impetus was related to security. Writing of the Ulster context, Philip Robinson suggests that settlers may have wished to congregate in the valleys, as these formed the points of entry and exit into the plantations through which the settlers could beat a retreat if they felt threatened by the native population.[9]

A third key point that figure 3.2 raises is the fact that it does not distinguish between the processes of organized colonization and longer-term informal population movement from Britain to Ireland. It has already been noted that modern-day Laois and Offaly in the east of Ireland were subject to the earliest wave of formal plantation under Mary I in the 1550s and that this was designed largely as a reactive measure in the defense of the Pale. In this sense the Pale is therefore an exceptional case where conventional plantation was not necessary, because colonial influence in this English cultural, economic, and religious bridgehead in Ireland was already strong. Prior to the Reformation, the Pale developed its hybrid identity as a zone of mixing between the English and Gaelic worlds.[10] By the nineteenth century the defenses may have come down, but the counties of Kildare, Meath, and Wicklow had morphed into the hinterland of a sophisticated colonial administration centered on Dublin. Dublin's pivotal role as capital and principal port of entry to the island meant that the entire region effectively

Fig. 3.3. The locations of evidence for Protestant populations in Munster, comparing the Elizabethan plantations, the 1641 depositions, and the 1834 Royal Commission. (Depositions data derived from N. Canny, "Early Modern Ireland c. 1500–1700," in *The Oxford Illustrated History of Ireland*, ed. R. F. Foster [Oxford: Oxford University Press, 1989], 135.)

Fig. 3.4. Church of Ireland and Presbyterian populations in 1834 overlaid by plantation precincts given over to English and Scottish undertakers.

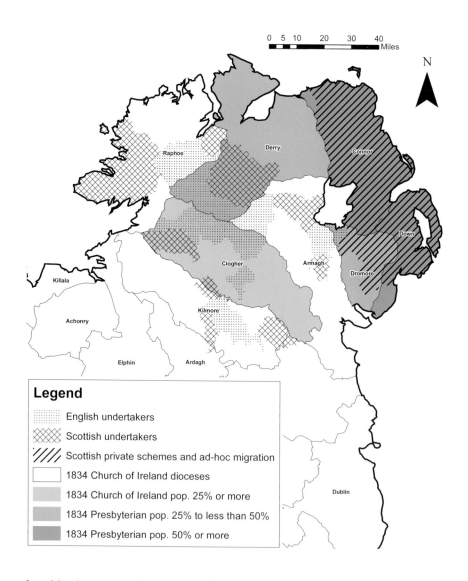

Legend

- ···· English undertakers
- ⊠⊠ Scottish undertakers
- ⧸⧸⧸ Scottish private schemes and ad-hoc migration
- ☐ 1834 Church of Ireland dioceses
- ▒ 1834 Church of Ireland pop. 25% or more
- ▒ 1834 Presbyterian pop. 25% to less than 50%
- ▓ 1834 Presbyterian pop. 50% or more

faced both ways and helps to explain the long-standing religious minority population on the central east coast.

Developing from this argument, the final point that emerges from comparison of the 1834 diocesan data with the original plantation patterns is the apparent marked success of such informal and ad hoc colonization schemes in Counties Antrim and Down over the more organized plantations that took place in the escheated counties in the west of Ulster. This is demonstrated in figure 3.4, which shows clearly that the two eastern counties that had large Protestant, indeed Presbyterian, populations were planted in an informal way. This may at first glance appear to be the ideological victory of private enterprise over government-led initiatives. In reality, the success of the east Ulster plantation was primarily the result of geography. Generations of Lowland Scots had been traveling over and back across the North Channel and had long-established trading and familial links with northeast Ulster. After the Reformation, which had taken a course in Scotland different from that in England, those Scots brought their own form of fundamentalist Protestantism with them.[11] The plantations formalized what had been, up to that point, a natural process of religious diversification in Ulster by specifically limiting the colonization scheme only to Scottish Lowlanders who had adopted Presbyterianism.[12] Scottish Highlanders had largely stayed true to Catholicism, and one of the main objectives of the entire Ulster plantation was to neutralize the enduring threat of rebellious alliances between dissident Catholics in Scotland and Ireland. Yet the Catholic Scottish Highlanders had also left their mark on Ulster's remarkable religious geographies; the Glens of Antrim remain to the present a strongly Catholic enclave surrounded by the overwhelming Protestantism of the rest of County Antrim, or, as A. T. Q. Stewart has defined it, "a Catholic bridgehead in the very heartland of Protestant territory."[13] Despite such anomalies, from the late sixteenth century many coastal areas of Antrim and Down culturally and religiously had more in common with the shores of Galloway and the rest of southern Scotland than with those of Lough Neagh. When the Scottish Lowlanders were encouraged to migrate on a larger-scale basis under the auspices of two Scottish nobles, James Hamilton and Hugh Montgomery, the existence of such strong trade and kinship ties in Antrim and Down enormously facilitated their transition.

It is true, then, that the Presbyterian population that settled in Antrim and Down and gave those counties their distinctively Scottish character did not necessarily face the same sort of challenges as the English and Scots who settled in the "frontier" counties west of the River Bann. This is borne out by analysis of the religious geography of the two main Protestant denominations now present in Ulster in large numbers. Figure 3.4 shows how the Church of Ireland and Presbyterian religious geographies reflect English and Scottish patterns of settlement in Ulster during the seventeenth-century plantation. The map also shows how the relatively dispersed Church of Ireland population failed to have the long-term demographic impact that the Presbyterian/Scottish population had in Antrim and Down. Once again, this is testament to the challenges faced by the Church of Ireland in western parts of Ulster in the early plantation period, but it also demonstrates the effects of scale and aggregation upon the analysis.

The long-term impact of private and ad hoc plantation in Antrim and Down is reinforced by the fact that these two counties effectively adhere to three of the 1834 dioceses. Thus, this concentration enabled the Presbyterian population to form over 50 percent of the entire population in two of the three eastern dioceses. However, although the Church of Ireland/English plantation pattern does appear to be relatively well concentrated in the midwest of Ulster, this does not follow through to the 1834 data, because the Church of Ireland/English are split between a number of territorial divisions that fall across the dioceses of Armagh, Clogher, Derry, Kilmore, and Raphoe. The only diocese where the Church of Ireland/English represent a significant minority is Clogher, and even there they are only 26 percent of the total.

Despite the limitations that the aggregate nature of the 1834 dioceses imposes, it is still evident that particular patterns of plantation settlement in Ulster in the sixteenth and seventeenth centuries had long-term impacts on religious geographies in the province that were still clearly discernible in the pre-Famine period. More specifically, the three main religions in Ireland had developed their own distinctive spatial patterns, which had been laid out in large part by the plantations but had by no means remained completely static over the two hundred years since formal colonization. By

Fig. 3.5. Population density of Irish baronies in 1821. Data have been mapped onto 1841 barony boundaries. Unshaded areas have no data due to boundary changes between these two dates.

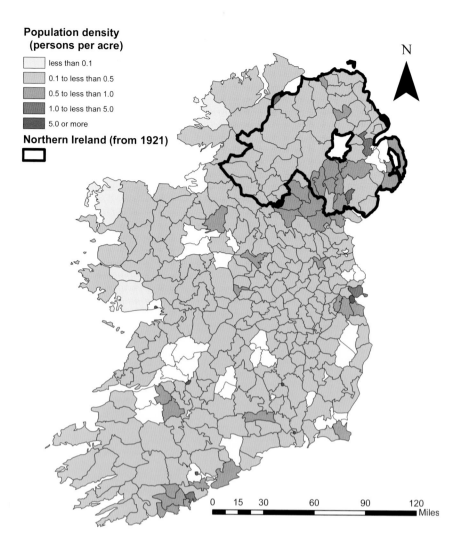

Population density (persons per acre)

- [ ] less than 0.1
- [ ] 0.1 to less than 0.5
- [ ] 0.5 to less than 1.0
- [ ] 1.0 to less than 5.0
- [ ] 5.0 or more

Northern Ireland (from 1921)

0   15   30   60   90   120
Miles

the time of the Famine the religious map of Ireland was a stark triptych reflecting the peculiar and partial impacts of colonization as a result of geopolitical imperatives.

Population

Having looked at how the religious geographies of the early nineteenth century were shaped by an earlier time, we now move to explore other aspects of Irish society in this pivotal period when Ireland began to move into the industrial age and immediately before it suffered from the last great famine in western Europe.

The 1821 census of Ireland returned a population of 6.8 million, whose distribution at the barony level is shown in figure 3.5. It shows that there were only a very few dense urban centers limited to places such as Dublin, Galway, Cork, Waterford, Drogheda, Limerick, and Kilkenny, all of which had population densities of over twenty persons per acre. Eastern and southern parts of Ulster, particularly Armagh, were relatively densely populated, with densities typically between 0.5 and 1 persons per acre. Parts of the Cork coast also had similar densities. Much of the remainder of Ireland, and particularly western and central areas, were, however, sparsely popu-

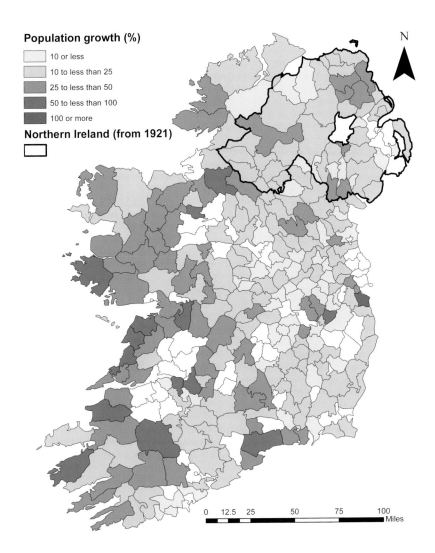

Population growth (%)

- 10 or less
- 10 to less than 25
- 25 to less than 50
- 50 to less than 100
- 100 or more

Northern Ireland (from 1921)

N

0 12.5 25   50   75   100
Miles

Fig. 3.6. Percentage of population growth by barony between 1821 and 1841.

lated, with the median barony population density being 0.30 persons per acre, and 82 percent of baronies having densities of less than 0.50.

This census was taken against the backdrop of a rapidly increasing population. Over the 1820s Ireland's population increased by 13 percent, and by 1841 it had reached almost 8.2 million, an increase of 21 percent over two decades. Ireland was not the only European country experiencing rapid population growth, but while much of the population growth in Britain and elsewhere was occurring in manufacturing towns and cities, in Ireland it was largely a rural phenomenon. While these early censuses do not provide evidence about agriculture, it is clear from other sources that population growth was fastest in those areas where farm holdings were smallest and the means of sustaining life most precarious.[14] In practice, as shown in figure 3.6, this generally meant western areas of the island.

Thus, despite rapid population increases, Ireland remained an over-whelmingly rural society in the pre-Famine period. Table 3.1 illustrates this by comparing Ireland to England and Wales, which were in the midst of the industrial revolution. People were flocking to the manufacturing centers emerging in the north and midlands of England and the valleys of south Wales. In 1831, 40 percent of the population of England and Wales lived in areas where the population density was in excess of one person per acre,

Table 3.1: Percentages of the population living at different population densities in Ireland and England & Wales in 1831.

Persons per acre	Ireland	England & Wales
More than 5.0	6.5	21.2
More than 1.0	8.5	40.0
0.5 to 1.0	25.1	9.9
Less than 0.5	66.4	50.1
Less than 0.1	0.4	1.2

Note that these figures must be treated with caution as they compare Irish baronies with hundreds and other similar units in England & Wales.

while 21 percent lived in districts with over five people per acre. The same figures for Ireland were just 9 percent and 7 percent, respectively, indicating that Ireland had not urbanized to anything like the same extent. Analysis of the more sparsely populated areas is also interesting. In England and Wales 50 percent of the total population lived in areas with a population density of less than 0.5 people per acre, and 1.2 percent lived where the density was less than 0.1 per acre. For Ireland the comparable figures are 66 percent and 0.4 percent. This tells us that a larger proportion of Irish people lived in rural areas; however, those living in the most sparsely populated areas were, relative to the entire population, fewer in number than in England and Wales. So while population growth in England and Wales was being soaked up by a relatively small number of very fast growing urban areas, population growth in Ireland was spread among a much larger number of primarily rural areas.[15]

Industrialization

As stated above, the pre-Famine censuses tell us nothing about agriculture. They do, however, provide information on industry, and this information provides some interesting clues as to why Ireland's population growth was so different from that of England and Wales. If urbanization is a product of industrial development and vice versa, then the census data suggest that Ireland was failing to industrialize, with the notable exception of the northeast. Figure 3.7 shows the percentage of the male population over the age of twenty employed in manufacturing in each county in 1831. Even at this date, Ulster and, more specifically, the east of the province were moving in an economic direction different from the rest of the island. In the twentieth century Belfast would become synonymous with heavy industry, most famously, shipbuilding, but even in the early nineteenth century Ulster was already the home of an established linen industry, which was dispersed particularly through Armagh and Down. This had demographic implications. As figure 3.5 shows, in the early part of the nineteenth century, Armagh was the most densely populated part of Ireland, and this density can be largely explained by the success of the linen industry.[16] However, mechanization was beginning to affect the industry, leading to an increased concentration of mills along the Lagan valley that used the most advanced steam-powered technology. This development was accelerating the processes of urbanization around Belfast.[17]

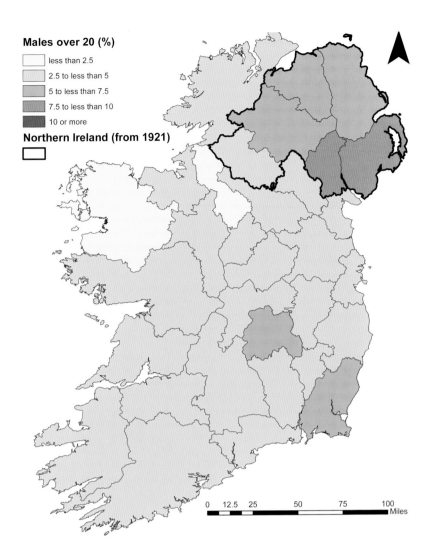

Males over 20 (%)

- less than 2.5
- 2.5 to less than 5
- 5 to less than 7.5
- 7.5 to less than 10
- 10 or more

Northern Ireland (from 1921)

0   12.5   25   50   75   100
Miles

Fig. 3.7. Percentage of the 1831 male population over the age of twenty employed in manufacturing by county.

The reinvigoration of the linen industry in Ulster as a result of technological advances in the 1820s produced a multiplier effect that worked through the wider economy in Ulster. This meant that the most northerly province was increasingly pulling apart from the rest of the island economically.[18] South Ireland's dependence on agriculture as its economic lynchpin had been encouraged by the boom years of the Napoleonic Wars. When peace returned to Europe after the Battle of Waterloo in 1815, it ushered in a period of economic depression in Ireland as the price of agricultural commodities fell. This was compounded by the removal of protective tariffs on Irish woolen and cotton goods in 1824, which saw the Irish market flooded with cheaper imports from Britain. This domestic industry had been one of the mainstays of the rural economy, particularly supporting those eking out a living on the marginal lands of the west. The liberalization of the sector dealt it a blow from which it never recovered.[19]

Emigration and Fertility

Emigration and fertility are two topics that have received a significant amount of academic attention, and this attention normally focuses on the impact of the Famine on these trends. It is clear, however, that in both cases behavior that is often attributed to the Famine was in fact in existence in the early nineteenth century. Increasing numbers of people started to emigrate from Ireland after 1815 as a response to the economic hardships of the time; however, even this emigration was merely the escalation of a long-established tradition.[20] Furthermore, it was not simply a characteristic of the increasingly impoverished and overpopulated western seaboard. Many migrants also left the more affluent regions of south Leinster and Ulster. Thus, while emigration reduced population pressure, it did not necessarily do so in areas where that pressure was most intense.[21] In western parts of Ulster, there also existed an established tradition of seasonal migration, most notably to the Scottish Lowlands to help with the annual harvest. Seasonal migration would increase in the latter half of the century and be a central part of the agricultural calendar in western parts of the province well into the twentieth century.[22] Nor was emigration an exclusively Catholic preserve; Protestants emigrated in numbers disproportionate to their share of the population and would join up with coreligionists from other parts of Ireland in North America or other parts of the New World to create their own distinctive communities.[23]

The orthodoxy on fertility is that the Famine precipitated significant and lasting changes in patterns of Irish sexual behavior, such as an increase in celibacy and the delaying of marriage. These patterns in large part relate to the system of inheritance in pre-Famine Ireland in which land was subdivided amongst children rather than following the rule of primogeniture. This, it is argued, led to unsustainable population pressure on finite and marginal land. However, Roy Foster has suggested that historians are no longer so comfortable with these orthodoxies and are rather more aware of our lack of knowledge of some of the complex cultural and socioeconomic dynamics at work in the run-up to the Famine.[24] Indeed, others, such as the economic historian Joseph Lee, have argued that trends toward later marriage and reduced fertility were in train well before the Famine and were "inherited" rather than "created" by the survivors.[25] Nevertheless, the subdivision of farms into increasingly uneconomic plots was one of the defining characteristics of pre-Famine rural society, particularly in the west of the island. In turn, the creation of smaller holdings with an ever-increasing population encouraged a growing dependence on a crop with the best yields and qualities that would allow it to flourish in the damp and windy climes of Connacht and west Munster: the potato.[26]

What is striking about nineteenth-century Ireland is the extent to which seemingly capricious aspects of social behavior such as marriage and fertility patterns became ruthlessly subjugated to economic imperatives. This would become particularly pronounced in the period after the Famine, with the Catholic Church getting a great deal of the blame for the repression of Irish sexuality. While this would not be without truth, the reality was that Catholic mores merely reflected the prevailing economic will of the broader population, and the prevailing will looked toward the consolidation of wealth through prudent marriage.[27] The Famine was not a watershed—the same dynamics were at play long before it occurred.[28] Once again, it is evident that those broader trends operated irrespective of religious persuasion if we take the example of the young Quaker quoted by C. Ó Gráda: "I am a single man still. I could have made a good match since I came to this town but my sweetheart found out that I was a weaver and would not have me. They know that weavers have not the means of keeping a wife and family, and no prudent wife would have one at all."[29]

Conclusions

This chapter describes the long-term effects of the plantations and the more immediate effects of the early industrial age on Ireland's geographies. The legacy of the plantations was the concentration of Presbyterians in northeast Ulster; a more dispersed distribution of the Church of Ireland in other parts of Ulster, east Leinster, and west Cork; and much of the rest of west and south Ireland being almost exclusively Catholic. Industrialization, in the form of the linen industry particularly, was becoming firmly established in Ulster, and technological trends were leading to this industry moving toward the Belfast area. This resulted in southeast Ulster already being the most densely populated part of the island. Population was growing, but Ireland was unusual in that this growth was most marked in rural areas, especially in the west, and while emigration and delayed fertility were already apparent, these were not sufficient to ameliorate the effects of this growth.

What will become apparent is that these two contrasting trends were beginning to coincide. Urban growth and industrialization were occurring in areas with large Protestant, especially Presbyterian, populations. Rural overpopulation was occurring in Catholic areas. When the potato blight struck, its impacts would have religious ramifications as well as social and economic ones.

4 The Famine and Its Impacts, 1840s to 1860s

It has almost become a cliché to argue that Ireland's population development over the last 150 years has been unique. It is the only developed nation in the world with a current population below that in the mid-nineteenth century and the only European country to have suffered a century of demographic decline in its recent history.[1] However, spatiopolitical qualifications must be applied to this assertion. The population decline of the area that is now the Republic has been remarkable, but the area that is now Northern Ireland was able to arrest its population decline at a much earlier stage. Furthermore, at the time of the Great Famine all of Ireland was part of the United Kingdom, and what might be described as a long-term regional population decline seems less spectacular when it is considered within the context of the U.K.'s rapid urban population growth, to which Irish migrants made a significant contribution.[2] Still, the impact of the Great Famine of the mid-nineteenth century on the shaping of modern Ireland cannot be trivialized. More than any other event it has defined both the literal and the metaphysical places of the Irish in the world. It has sent shock waves down through the centuries that are not only demographic but also socioeconomic, cultural, and political.

An Gorta Mór—the Great Hunger

When the potato crop first became infected with blight in the autumn of 1845, it did not signify an unprecedented event; there had been previous crop failures, and there had been famines as a result of these failures.[3] What turned a crisis into a catastrophe were the peculiar circumstances that pertained in Ireland. For three million people the potato was the food on which they depended for survival—the average male laborer consumed up to 14 lb. (6.3 kg) of the tuber per day.[4] It is easy to see why such a huge mass of the Irish population became dependent on the potato. It was a hardy and high-yielding crop, perfectly suited to Irish climatic conditions and economic circumstances. The dramatic increase in the rural population described in chapter 3 meant that farm sizes decreased through subdivision, resulting in the potato becoming the best means of sustaining life on a wet and windy marginal holding.[5] More recent research shows the nutritional advantages of the potato, and contemporary commentary on the physical appearance and general health of the rural poor appears to support this research.[6] Joel Mokyr's statistical analysis of the background to the Famine has found that dependence on the potato actually had a positive effect on the high rates of Irish fecundity in the pre-Famine period.[7] The British government failed to foresee the impending disaster, and the limited efforts that the government made to counter that disaster's impact proved impotent. It is also likely that previous smaller crises that Ireland had weathered had led to a sense of complacency among the highest levels of the central administration.[8] From the point of view of the pre-Famine Irish peasantry, within their limited terms of economic reference they were behaving in a perfectly logical manner in opting to exploit the one food resource capable of sustaining a rapidly growing population on a finite and difficult terrain.

In 1845 only about half the harvest had actually been lost. Had this been an isolated incident, Ireland might have come through the winter with relatively few excess deaths. However, the almost complete collapse of the 1846 crop signaled the start of what was to become a crisis of unparalleled proportions. During the severe winter of 1846–47 the death toll began to climb rapidly. Hunger was compounded by rampant disease brought on by lowered levels of immunity. "Famine fever," a combination of typhus and relapsing fever, dysentery, and its more virulent form known as the "bloody flux" spread through the population.[9] Even scurvy, a result of vitamin C deficiency, which had been largely unheard-of prior to the Famine, became rife among those who were compelled to eat the maize imported by the British government at the Famine's outset.[10]

In 1847 only a fraction of the usual acreage of potatoes was planted because in desperation people had resorted to eating the seed potatoes that were essential for the following year's harvest. This resulted in a collapse in the cultivated area, which fell from an estimated two million acres in 1845 to less than three hundred thousand just two years later.[11] However, 1847 realized high yields for the relatively tiny acreage sown that year, and the following year cultivation levels rallied in the expectation or hope that the worst had passed. It was a cruel hoax, as the 1848 crop once again failed and brought the death toll higher still.[12] The crop gradually recovered in subsequent seasons, but by the end of 1851 almost 1.1 million people are believed to have died directly as a result of the Famine.[13]

Demographic Impacts

Historians have traditionally had some difficulty in dealing with the Famine. In the mid-1990s the 150th anniversary of the disaster brought renewed academic interest and a flurry of new publications, yet prior to that the response of historians was largely not to deal with the Famine at all.[14] There are understandable reasons for that reticence. First, the Famine is, in an Irish context, unparalleled in the scale of human trauma it represents and, concomitantly, in its potential to arouse conflicting opinions and high emotions. The Ulster Presbyterian and revolutionary nationalist John Mitchel powerfully accused the British government of genocide soon after the Famine, and, indeed, the respected English historian A. J. P. Taylor made the same charge almost exactly a century later.[15] While the British have been cleared by most historians under that indictment, the fact that Prime Minister Tony Blair in 1998 expressed regret for British actions—or, more accurately, inaction—during the Famine was testament to a profound belief in the power of the historical imagination to explain the course of

Anglo-Irish relations since the Famine.[16] When Foster states that the Famine left not only a legacy of demographic decline, altered farming, and economic structures but also "an institutionalized Anglophobia among the Irish at home and abroad," the observation is not without truth, but it tells us more about the present in which Foster's book was written rather than the past.[17] A historiographical vacuum has only recently been filled because non-native Irish scholars have taken the debate above the parochial concerns of whether or not their findings give ideological succor to political extremists in Northern Ireland and beyond.[18]

The second historical explanation for an academic reluctance to deal with the Famine lies in the fact that there has been an increased awareness of what is unknown about the Famine.[19] A lack of statistical and other evidence means that historians were simply less confident about dealing with the issue than perhaps is the case now with the benefit of new research over the last fifteen years or so. Relating to this is an increased sensitivity to the spatial variation that appears evident within the impact of the Famine, making historians more reluctant to make general claims about its impact. In this sense, L. M. Cullen's controversial claim that the Famine was more of a "regional" disaster than a "national" one is worthy of further attention here.[20]

Figure 4.1 shows Ireland's population density in 1841 and 1861. Comparing the 1841 pattern with that of 1821 (shown in figure 3.5) shows the continuing population growth over this period, which contrasts with the decline from 1841 to 1861. It is obvious that most population change during the period can be explained by deaths and emigration resulting from the Famine. It is difficult to gauge precisely where emigration rates were highest, but J. R. Donnelly has identified south Ulster, north Connacht, and the Leinster midlands as having particularly high levels.[21] Furthermore, that emigration was not a short-term response to the Famine but rather part of a longer-term trend reflecting a semi-permanent economic crisis in rural Ireland.[22] This relates back to the growing academic awareness that so many population dynamics in post-Famine Ireland had their roots much farther back in the century rather than simply emerging in the Famine years. In this sense the Famine was not a watershed in modern Irish history. If there was a watershed, it can be argued to have been Waterloo in 1815 and the prolonged depression that it ushered in.[23]

Although the barony units vary widely in area, and many of these larger units are in the west and north of the island, in the period up to 1841 the pattern of greatest population growth occurred across mid-Ulster and large tracts of the south and west of Ireland (see figure 3.6). Figure 4.1 shows that by 1861 it was apparent that the population had fallen rapidly across many areas but with the decline being particularly pronounced in Connacht and Munster. In stark contrast to this trend, the historically high population levels across the center and east of Ulster remained high. Mokyr has calculated the excess death rate per county during the Famine years, excess deaths being those above what would normally be observed and therefore highly attributable to the Famine. Mokyr's lower-bound estimates for excess deaths highlight the remarkably intense spatial variation in the Famine's impact. The five counties with the highest rates are those that make up the province of Connacht, Mayo having the highest level of death, with an

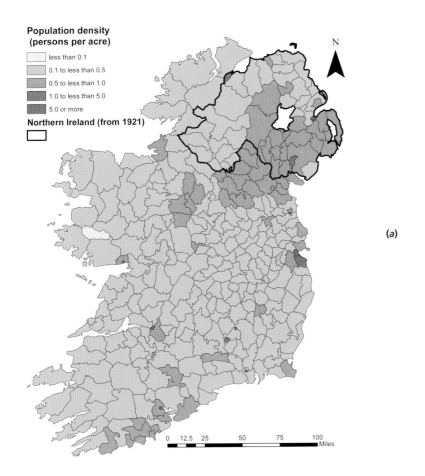

(a)

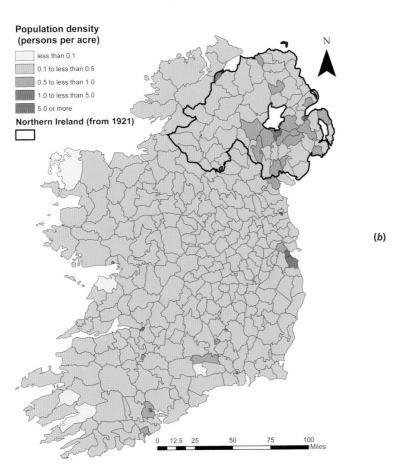

(b)

Fig. 4.1. Population density of Irish baronies in (a) 1841 and (b) 1861.

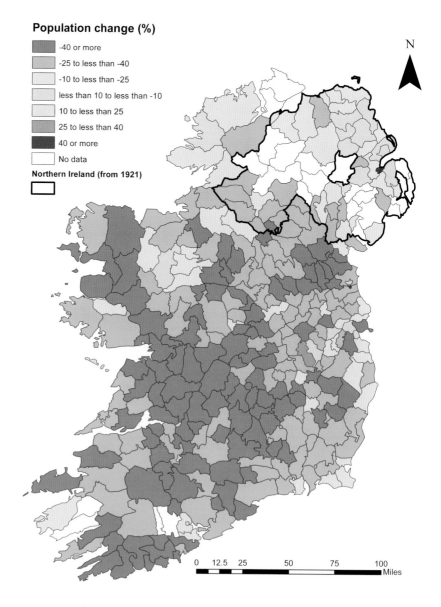

Population change (%)

- -40 or more
- -25 to less than -40
- -10 to less than -25
- less than 10 to less than -10
- 10 to less than 25
- 25 to less than 40
- 40 or more
- No data

Northern Ireland (from 1921)

N

0   12.5   25   50   75   100
Miles

excess mortality rate of 58.4 per thousand.[24] It is clear from figure 4.1 that Mayo saw exceptional falls in its population during the twenty years from 1841 to 1861. Yet the maps show that even in this most devastated county the pattern was not uniform, with particularly high populations being enumerated in the baronies of Tirawley and Carra. Overall, 40.4 percent of excess deaths occurred in Connacht, 30.3 percent in Munster, and 20.7 and 8.6 percent in Ulster and Leinster, respectively.[25]

Figure 4.2 encapsulates the period of most dramatic change by summarizing at the barony level where the population fell or increased between 1841 and 1861. The map shows that while some coastal areas along the western seaboard saw substantial falls in population, particularly in Mayo, west Kerry, and southwest Cork, many inland baronies stretching through north Munster as well as north Leinster also saw significant falls in population. These are areas that would not traditionally be associated with the most severe impacts of the Famine. A number of baronies in Ulster were split between the two dates; therefore, it is impossible to directly compare the data for 1841 and 1861, as the geographical areas are not consistent. The map does indicate that larger parts of Ulster saw population stability in the twenty-year period spanning the disaster and that Belfast in particular saw rapid growth. The city was split between two baronies, one in Down and one in Antrim, that experienced population increases of 44 and 63 percent, respectively. Among other areas, Drogheda grew rapidly, while Dublin and Waterford saw more modest growth. Cork and Galway remained relatively stable, while Limerick City fell by 8 percent. Thus, the pervasive notion that the Famine automatically led to the increasing urbanization of Irish society needs to be qualified. Some urban centers did increase during this period, but those that did tended to reflect the spatially varying impacts of the Famine on the hinterland population. The major increase in the population of Belfast was indicative not only of the path of rapid industrial expansion, which was by then well advanced, but also of the fact that the Famine appears to have had a relatively low impact in east Ulster. Cork, Galway, and Limerick, on the other hand, were surrounded by hinterlands that suffered major population slumps between 1841 and 1861. This decline appears not to have led to a significant rural migration into these towns. Instead, the decimation of the surrounding population appears to have inhibited their potential to grow in the short term.

Another reason that is likely to have militated against centralized urban growth in a few centers was the development of a network of workhouses to cater to the starving and destitute. These workhouses soon proved desperately inadequate to deal with the level of demand, but the subdivision of the country into 163 Poor Law Unions by 1850, each with a workhouse, tended to deflect the stricken toward smaller urban centers despite the appalling and well-founded reputation that the workhouses had.[26] Before qualifying for entry, a "pauper" had to be in a state of "total destitution," and this meant surrendering any landholding of more than one-quarter of an acre.[27] Once inside, families were broken up and subjected to a discipline and work regime designed to make life inside the workhouse as unpalatable as possible. The workhouses acted as localized lightning rods not only for destitution but for disease as well. Overcrowding and woefully inadequate facilities meant that those suffering from hunger-related diseases such as typhus, dysentery, and relapsing fever were often mixed up with the "healthy" inmates, with appalling consequences. In the Fermoy workhouse this situation resulted in a 24 percent mortality rate in the winter of 1846–47, yet, despite the fearsome reputation of the workhouses, such was the level of desperation that the one hundred thousand places in the workhouse system were oversubscribed by a rate of 5:1 during that season.[28]

The unusual population growth in the barony of Offaly East in County Kildare is likely to be explained by its strategic importance. The barony was (and, indeed, remains) home to the largest military installation in Ireland, known as the Curragh Camp. In 1848 members of the Young Ireland movement rebelled, distilling the revolutionary impulse sweeping Europe in that year as well as indigenous grievances against the British.[29] While the rising was a dismal failure, contemptuously dismissed by the *Times* of London as a "cabbage-garden revolution," it did underline the continued threat of instability that Ireland represented, particularly against the social and economic upheavals of the Famine.[30] It is highly likely, therefore, that population growth in this barony can be explained by the reinforcement of British troops at the main infantry base on the Curragh.

Fig. 4.3. Ireland's three largest religions by barony in 1861 showing (a) Catholics, (b) Church of Ireland, and (c) Presbyterians.

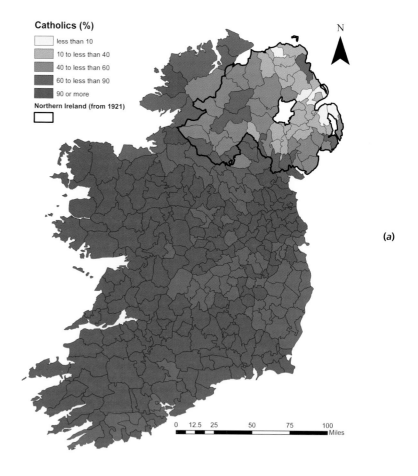

Catholics (%)

less than 10
10 to less than 40
40 to less than 60
60 to less than 90
90 or more

Northern Ireland (from 1921)

N

(a)

0 12.5 25 50 75 100
Miles

Religious Change

Figure 4.3 shows Ireland's three major religions by barony in 1861. The maps indicate a more detailed pattern of religious settlement, but one that is entirely consistent with that observed from the 1834 data shown in figure 3.1. Catholics predominate across all of Munster, Connacht, and most of Leinster, gradually declining in proportion from west to east across Ulster. For the Church of Ireland, the pattern is across south Ulster from the North Channel to the Atlantic. Some pockets of the established church also stand out in Leinster and Munster, consistent with the plantation patterns and areas of long-standing English/British influence in Ireland around the Pale. The Presbyterian population in 1861 was heavily concentrated in the northeast of Ulster, in the heartland areas of Antrim and Down. Enumerated separately, the two Protestant denominations have never represented the sort of overwhelming majority of the population that the Catholics have had. However, in some areas such as north Armagh, north Down, and south Antrim there is considerable overlap between the two denominations, and in these areas, when considered together, Protestants represented a substantial majority of the population.

The first year in which religion formed part of the population census was 1861, and for that reason, the census results were eagerly anticipated by all parties anxious to have statistical evidence to support their own political agendas. For Catholics and Presbyterians, and indeed for many English liberals, the central part of their agenda was the disestablishment

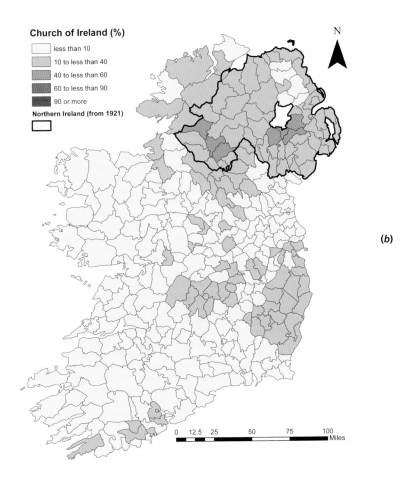

(b)

N

0 12.5 25 50 75 100
Miles

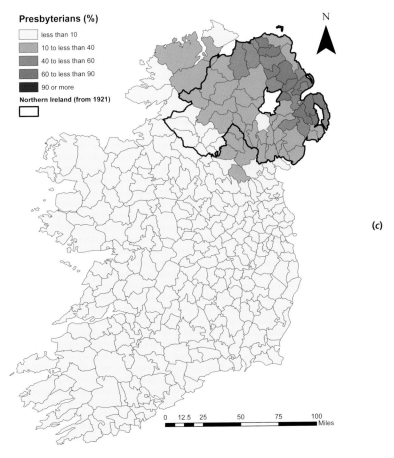

(c)

N

0 12.5 25 50 75 100
Miles

The Famine and Its Impacts, 1840s to 1860s   43

Fig. 4.4a and 4.4b. The change in population and of religions by diocese between 1834 and 1861, showing (a) total population, (b) Catholics. (*Continues on the next page*)

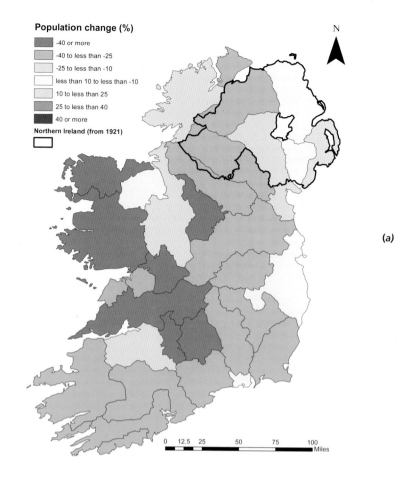

Population change (%)

- -40 or more
- -40 to less than -25
- -25 to less than -10
- less than 10 to less than -10
- 10 to less than 25
- 25 to less than 40
- 40 or more

Northern Ireland (from 1921)

0 12.5 25 50 75 100 Miles

(a)

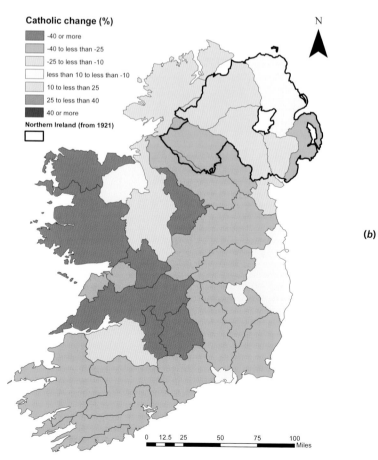

Catholic change (%)

- -40 or more
- -40 to less than -25
- -25 to less than -10
- less than 10 to less than -10
- 10 to less than 25
- 25 to less than 40
- 40 or more

Northern Ireland (from 1921)

0 12.5 25 50 75 100 Miles

(b)

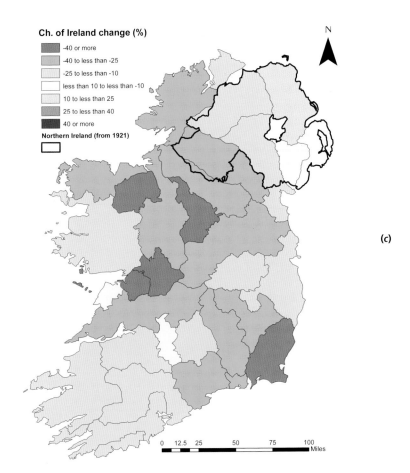

Ch. of Ireland change (%)

- -40 or more
- -40 to less than -25
- -25 to less than -10
- less than 10 to less than -10
- 10 to less than 25
- 25 to less than 40
- 40 or more

Northern Ireland (from 1921)

N

(c)

0   12.5   25   50   75   100
                                        Miles

Fig. 4.4c and 4.4d. The change in population and of religions by diocese between 1834 and 1861, showing (c) Church of Ireland, and (d) Presbyterians.

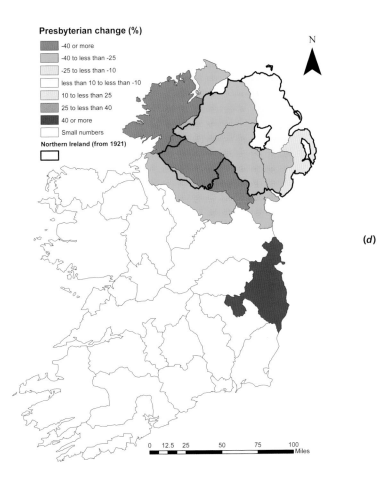

Presbyterian change (%)

- -40 or more
- -40 to less than -25
- -25 to less than -10
- less than 10 to less than -10
- 10 to less than 25
- 25 to less than 40
- 40 or more
- Small numbers

Northern Ireland (from 1921)

N

(d)

0   12.5   25   50   75   100
                                        Miles

of the Church of Ireland. Catholics and Presbyterians resented the privileged position of the established church, which they felt was based on an inflated and false supposition of its demographic weight. If the census returns showed that membership of the Church of Ireland was significantly lower than had previously been thought, this would lend support to calls for its disestablishment, and this is precisely what happened just eight years later.[31] Those on the more evangelical wing of Presbyterianism optimistically hoped that the census would reveal that Protestants and not Catholics were in fact the majority population in Ireland.[32] They were to be roundly disappointed in that aspiration, because the 1861 enumeration showed that Catholics still made up about 78 percent of the population.

Comparing the 1834 Commission data with the 1861 census allows us to explore how Ireland's religions were affected by the Famine. The overall population loss between these two dates was 27 percent, or 2.15 million people. Catholics were the most severely affected of the religions, with their numbers declining by 30.4 percent, compared to 20.7 percent for the Church of Ireland and 18.8 percent for the Presbyterians. A rise in the number of "other Protestants" actually means that the overall decline in the Protestant population was 16.2 percent. Overall, therefore, the Catholic population was more severely affected than the Protestant population in proportional terms, and, given their numerical superiority, this meant that the Catholics' population loss was almost eight times higher than the Protestants' loss.

Interpolation allows us to compare the geographies of these losses at the diocesan level. Figure 4.4 shows the percentage change of each religion over this period together with overall population change at the diocese level for reference. The maps show that population loss was concentrated in the west and midlands, where many dioceses experienced population declines of over 40 percent in the seventeen-year period. Connor (which approximates to Antrim and includes Belfast) was the only diocese to experience a gain in population, and this was only by 6.6 percent. The change in the distribution of the Catholic population mirrors this pattern well, with large losses in the west and midlands and smaller losses elsewhere. Connor showed an increase in its Catholic population of 9 percent, while Achonry, in northeast Connaught, also showed a modest gain of 1.1 percent. The pattern of loss of the Church of Ireland also shows major losses across much of Ireland, with the exceptions again being the northeast of Ulster, where the Church of Ireland's population increased by 20.4 percent. Presbyterians are harder to map, as their numbers were so small across much of Ireland. As a consequence, dioceses with fewer than 1,000 Presbyterians in 1834 have not been shaded. Even with this precaution, the apparent large rise in Dublin is only from 2,300 to 6,500. Apart from this, the pattern is largely similar to that of the other religions, with widespread losses except in the northeast of Ulster. Even in Connor the Presbyterian population fell, albeit by only 4.2 percent.

It is clear that Catholics were disproportionately affected throughout the Famine period; however, this occurred because, as is shown in figure 4.5, the Famine was most severe in areas where the population was overwhelmingly Catholic. It is interesting that in more mixed areas the decline in Protestants seems to have been at least as severe as the decline in Catholics. The result of this is that while the Famine led to the overall

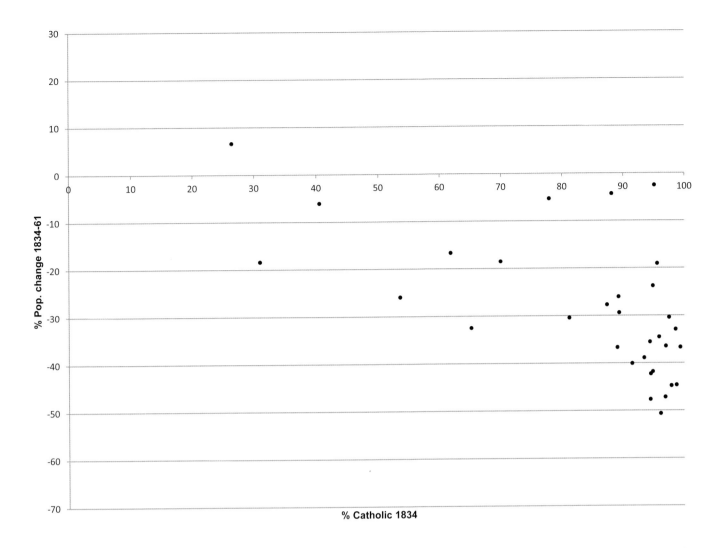

Fig. 4.5. The relationship between the Catholic population in 1834 and overall population loss between 1834 and 1861. The x-axis shows the percentage of the population that was Catholic in 1834; the y-axis shows total population loss between 1834 and 1861. Data have been interpolated onto Church of Ireland dioceses.

Catholic majority of the population falling from 81 percent to 77 percent, the religious geographies did not change much. This is shown in figure 4.6, which compares the percentage of the population that was Catholic in 1834 with 1861. In general the overall percentages have changed very little. The only area that became significantly more Catholic was Raphoe (Donegal), where the Catholic proportion of the population grew from 70 to 77.2 percent, while Cashel (96.9 to 89.9), Emly (98.7 to 93.6), Down (31 to 26.1), Cork (89.3 to 84.4), and Ardagh (91.5 to 87.1) became less Catholic. Of these, perhaps Down is the most interesting, as it was already one of the most Protestant parts of Ireland and became more so over the Famine; however, its neighbor Connor (Antrim) showed very little change, rising from 26.4 to 27 percent Catholic.

Social segregation can be measured using a number of quantitative indicators. The index of dissimilarity attempts to measure the proportion of the population that would have to move to allow the population to be evenly distributed between all districts.[33] Population changes over the Famine period resulted in the index of dissimilarity between Catholics and Protestants falling from 54.3 percent in 1834 to 50 percent in 1861. This relatively large change probably reflects the overall loss of Catholic population: fewer members of a smaller group would have to move in order to evenly distribute that group across the entire population. An alternative measure

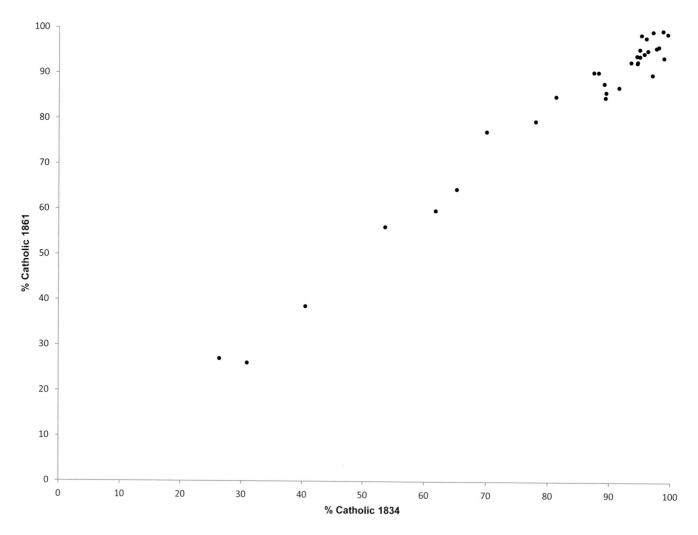

Fig. 4.6. The relationship between the Catholic population in 1834 and 1861. Data have been interpolated onto Church of Ireland dioceses.

of social segregation is the isolation index, which attempts to measure the chance of a person of one group interacting with another member of the same group if that person is selected at random from the population.[34] If, for example, 80 percent of the population is Catholic, then we would expect there to be an 80 percent chance of a Catholic interacting with another Catholic, assuming that Catholics were evenly distributed across Ireland. As the actual value rises above an expected figure, it indicates that the group is isolated from interacting with members of another group. The isolation index for Catholics fell from 80.7 percent to 77.2 percent, but this is not as much as would be expected given the overall loss of Catholic population. The difference between the actual and the expected isolation index rose from 5.4 to 7 percentage points. For Protestants, the isolation index rose from 41.7 percent to 42.1 percent, but this is not as much as would have been expected given the proportional rise in the Protestant population, so the difference between the actual and the expected fell from 22.9 percent to 21.4 percent. This slightly confused picture therefore seems to confirm that the Famine did not have a major impact on changing the religious geography of Ireland except for the disproportionately large loss of population from the overwhelmingly Catholic parts of west and south Ireland.

The fact that the Famine disproportionately impacted Catholic areas or, perhaps more accurately, that Catholics were overrepresented in the poorest areas, which were, in turn, hardest hit by the Famine, served to reinforce sectarian divisions in the post-Famine decades.[35] Such tensions were no doubt exacerbated by a particular strain of evangelical thought; the Famine's spatial impact inspired that strain's believers to check all the right sectarian boxes of Protestant providentialism, which saw the hand of God in all events. In the historiography of the Famine, much of the blame has been laid squarely at the door of the treasury secretary, Charles Trevelyan. Historian Jenifer Hart suggests that he saw the disaster as "the judgment of God on an indolent and unself-reliant people, and as God has sent the calamity to teach the Irish a lesson, that calamity cannot be too much mitigated: the selfish and indolent must learn their lesson so that a new and improved state of affairs must arise."[36] More recent scholarship has questioned the accuracy of this characterization and suggests that Trevelyan has in fact been unfairly demonized in successive histories of the period.[37] Certainly, he maintained both a profound belief in individual self-reliance and a strong theological objection to Catholicism, but this was not reflective of a deeper hostility to the generalized mass of the Irish population.[38] In many respects his views on the Famine were simply the articulation of more commonly held beliefs about the disaster and its causes.[39]

Social Change

While Trevelyan, according to Hart's reading, employed ethnonational modes of contextualizing Ireland's economic problems, many commentators in the pre-Famine period, and indeed more recently, were convinced that overpopulation in Ireland meant that such a disaster was inevitable. This is the root of the "Malthusian apocalypse" hypothesis, which greatly influenced economics in the early nineteenth century. One of its more influential proponents was the Oxford academic Nassau William Senior, who accurately guessed the overall death toll of one million while commenting that it would scarcely be enough to do much good.[40] Mokyr's recent work has done much to dismiss the Malthusian notion that Ireland's ills were based solely on overpopulation, while Peter Solar has argued that the Great Famine was unique in so many respects that glib generalizations are simply not helpful in understanding the scale of its impact.[41] Much harder to dismiss, however, is the substance of Senior's analysis: that a significant depopulation of the island would lead to an improvement in living standards for those who remained. It is a view that has been endorsed by modern economic historians such as Ó Gráda and is borne out by an analysis of the following key indicators.[42]

Figure 4.7 shows the percentage of each barony population that could not read or write in English between 1841 and 1861. In 1841 those illiterate in English were heavily concentrated on the western seaboard, with baronies covering all of Connemara in western County Galway and areas of Mayo, south Kerry, and the Ring district of County Waterford in the 80 percent or above category. Through the 1851 and 1861 censuses a clear pattern of contraction is discernible as the population illiterate in English declines. The area with the lowest levels of illiteracy in English are those around Belfast, with many baronies in Down and Antrim below 20 percent. These

Fig. 4.7. Illiteracy in English by barony:
(a) 1841, (b) 1851, and (c) 1861.

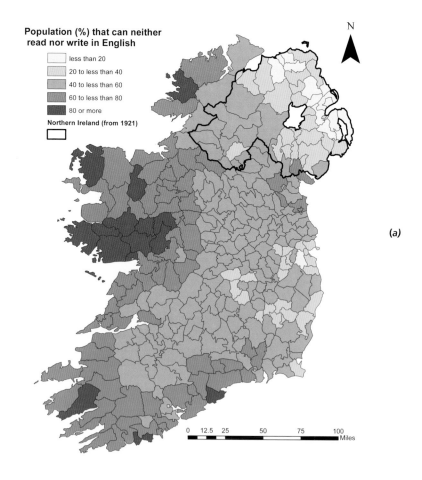

Population (%) that can neither
read nor write in English

- less than 20
- 20 to less than 40
- 40 to less than 60
- 60 to less than 80
- 80 or more

Northern Ireland (from 1921)

N

0   12.5   25   50   75   100
                                        Miles

(a)

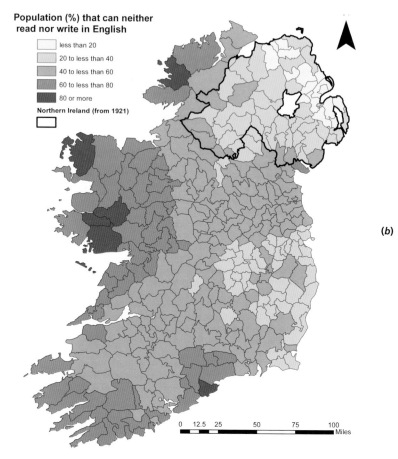

Population (%) that can neither
read nor write in English

- less than 20
- 20 to less than 40
- 40 to less than 60
- 60 to less than 80
- 80 or more

Northern Ireland (from 1921)

0   12.5   25   50   75   100
                                        Miles

(b)

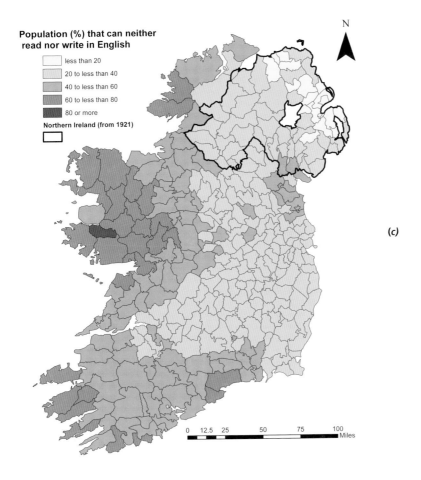

Population (%) that can neither read nor write in English

less than 20
20 to less than 40
40 to less than 60
60 to less than 80
80 or more
Northern Ireland (from 1921)

N

(c)

0 12.5 25 50 75 100 Miles

figures must be interpreted with care, as it is unclear whether they measure low educational attainment or the use of the Gaelic language.

Figure 4.8 gives the percentage of all housing defined as fourth class in the censuses between 1841 and 1861. Fourth-class housing, comprised of "mud cabins having only one room," was the lowest caliber of housing on the census. Kennedy and colleagues have queried how consistently the various classifications were enforced across the country, but it is reasonable to assume that such dwellings, even by the modest standards of the day, were of exceptionally poor quality.[43] The map for 1841 shows that such houses were overwhelmingly concentrated in the western half of the island on the eve of the Famine. The following two maps in the series indicate a remarkable decline in the poorest quality of housing over the ensuing twenty-year period. By 1861 only the entire county of Kerry and isolated parts of west Cork, Galway, and Mayo show concentrations of poor-quality housing of 20 percent or above.

These two analyses would appear to be a perfect vindication of Senior's prediction, discussed above, that the Famine would precipitate a rise in living standards for those who remained. This is largely true, but the cost in human terms was incalculable. These maps testify not only to the influence of philosophies of laissez-faire political economy over Ireland's development during the Famine but also, more precisely, as even Senior himself observed, to the unfettered manner in which Ireland was used as a laboratory for the testing of such doctrines.[44] The reason why illiteracy in English declined and housing quality improved was because that entire

Fig. 4.8. Fourth-class housing as a percentage of all dwellings per barony: (*a*) 1841, (*b*) 1851, and (*c*) 1861.

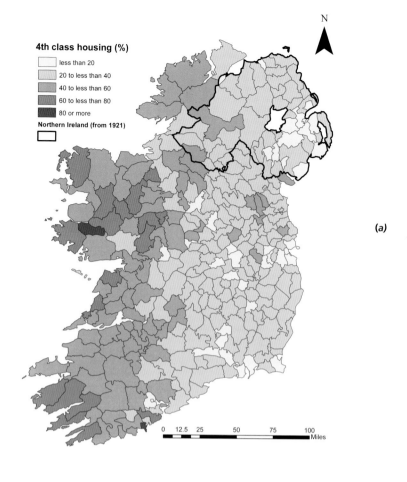

4th class housing (%)

- less than 20
- 20 to less than 40
- 40 to less than 60
- 60 to less than 80
- 80 or more

Northern Ireland (from 1921)

0 12.5 25 50 75 100
Miles

(*a*)

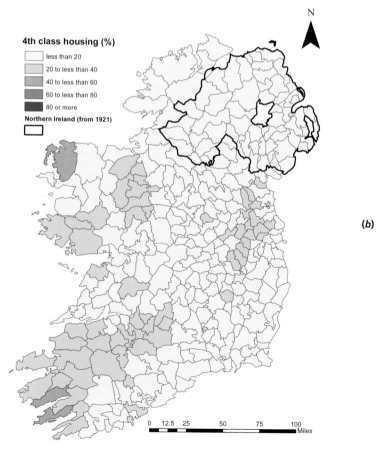

4th class housing (%)

- less than 20
- 20 to less than 40
- 40 to less than 60
- 60 to less than 80
- 80 or more

Northern Ireland (from 1921)

0 12.5 25 50 75 100
Miles

(*b*)

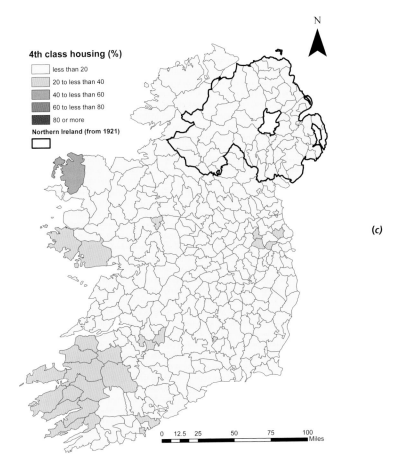

4th class housing (%)
- less than 20
- 20 to less than 40
- 40 to less than 60
- 60 to less than 80
- 80 or more

Northern Ireland (from 1921)

N

(c)

0 12.5 25   50   75   100
                                Miles

class of people who spoke Gaelic and lived in fourth-class housing—the landless laborers, cottiers, and small farmers—disappeared from the Irish landscape between 1841 and 1861.[45] It was a heavy price to pay for such an improvement.

Economic Impacts

In economic terms the Famine was perhaps more remarkable for its continuities rather than its disruptions, which can in part be explained by its spatial impact. We have established that the Famine disproportionately affected the west of the country and the poorest tier of the population. The economic impact of the rural poor did not reflect their demographic strength, as so many of this tier were living at or near subsistence level. This group numbered some three million on the eve of the Famine, most surviving on combined annual family incomes of £15 to £20 a year, leading Ó Gráda to estimate that the poorest 40 percent of the Irish population received only 10 to 15 percent of the entire national income.[46] That constituted £9 million to £12 million, of which the same amount was paid back in annual rents to just ten thousand landlords.[47]

Catastrophic as it was, the Famine did present an opportunity for the radical remolding of the Irish economy. That opportunity was not taken. Politicians such as Chancellor Charles Wood were ideologically opposed to pouring more money into the Irish economy after the relief efforts of the Famine and preferred to wait for private enterprise to take up the slack. However, the anticipated injection of capital did not materialize, and so im-

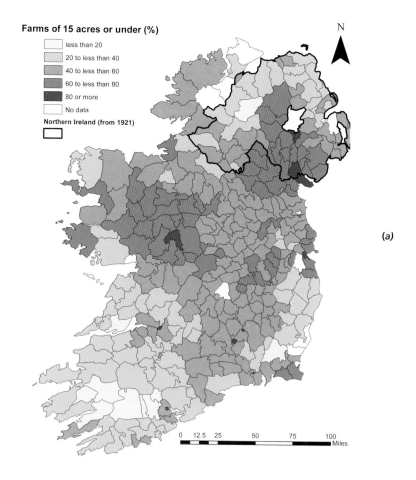

Farms of 15 acres or under (%)

- less than 20
- 20 to less than 40
- 40 to less than 60
- 60 to less than 80
- 80 or more
- No data

Northern Ireland (from 1921)

(a)

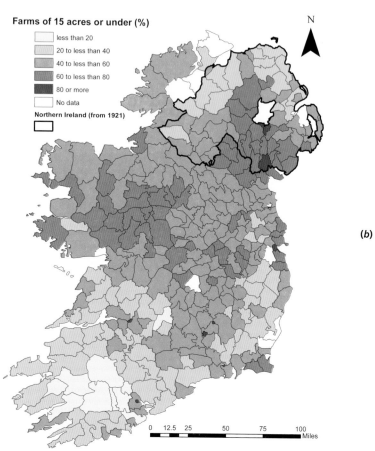

Farms of 15 acres or under (%)

- less than 20
- 20 to less than 40
- 40 to less than 60
- 60 to less than 80
- 80 or more
- No data

Northern Ireland (from 1921)

(b)

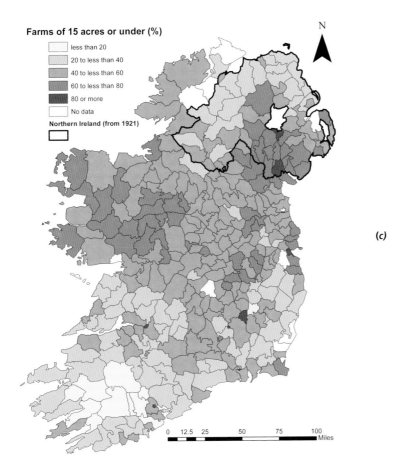

Farms of 15 acres or under (%)

less than 20
20 to less than 40
40 to less than 60
60 to less than 80
80 or more
No data

Northern Ireland (from 1921)

N

(c)

0   12.5   25   50   75   100
                                    Miles

provements in the agricultural sector were retarded.[48] In many respects the Famine was a recognition of the failure of a rural economy based heavily on iniquity and, outside northeast Ulster, almost completely on agriculture. In the next chapter, the reasons for Ireland's limited industrial development will be discussed in more detail, but clearly one of the key reasons why the northeast of the island was less directly affected by the disaster was its diverse economic base.[49] The pervasive influence of laissez-faire economic thinking, which advocated limited government interference with free-market dynamics, helped to explain not only the government's unwillingness to manage the direction of the Irish economy but also its entire response to the Famine crisis.[50]

An agricultural system in large part characterized by inequality and inertia thus contextualizes the pattern that can be seen from figure 4.9, which shows baronies with the highest concentrations of small farms, those of 15 acres or less. Throughout this time frame there is some contraction, although the overall pattern remains relatively stable. The areas with the highest proportions are in east Connacht and the north midlands as well as south Ulster. The concentration of smallholdings in south Down, Armagh, and east Tyrone are partly explained by the historically high population densities in these areas, which was in turn due to the proximity of these flax-growing regions to the insatiable linen mills of the Lagan valley.[51] The cluster of baronies in the north midlands/east Connacht region is indicative of the impact of the Famine and the fact that larger population loss farther west had enabled greater consolidation of smallholdings into larger and more economically viable farms.[52]

The Famine and Its Impacts, 1840s to 1860s   55

These developments suggest that a degree of entrepreneurship was creeping into the Irish agricultural system, and this is not without some validity. The Famine swept away the subsistence system, and a new class of small farmers was to ascend in Ireland. These farmers had always been present and had not simply survived the Famine unscathed but consolidated their position considerably as a result of it.[53] They showed their economic prudence by embracing the market system and shifting from tillage to livestock to take advantage of transport developments that had improved access to the British market.[54] The increase of material wealth directly influenced social changes, which were touched on in chapter 3, such as an increase in celibacy and the delaying of, and abstention from, marriage.[55]

So underlying the apparent continuities in the agricultural system suggested by figure 4.9 was a high degree of economic reactivity and innovation. On the other hand, the continuities within the maps reflect the reality that real power and influence still lay not with these consolidating farmers but with the landlords. The consolidating farmers showed a high degree of responsiveness to a changing economic structure, but their terms of reference were still limited by the inequalities of the land system in Ireland.[56] The landlords saw their incomes collapse during the Famine and, by resorting to evicting tenants at its height, created a legacy of bitterness that would last for generations.[57] The double impact of falling incomes and crippling poor law rates led to the bankruptcy of at least 10 percent of the demesnes, which were then sold off under the Encumbered Estates Act of 1849.[58] The track records of individual landlords varied, many taking an active and progressive interest in their estates. However, as a group, they were not highly regarded on either side of the Irish Sea and can be seen to have contributed to their own downfall through a mix of mismanagement, greed, and absence.[59] Nothing radically changed immediately—the estates were not broken up among the local landless natives who remained, for example. However, the vaunted position of the landlord class had been permanently diminished by the impact of the Famine. The Famine had put land reform on the agenda, and it would become the supreme political issue of late nineteenth-century Ireland.[60]

Figure 4.10 builds on figure 3.7 to show the male population employed in the manufacturing sector as a percentage of all males in each county. As figure 3.7 showed, in 1831 most of the counties of Ulster and Leinster gradually pulled away from the rest of the south and west of the country in terms of manufacturing employment. However, the differences were not yet substantial. The heaviest concentration of manufacturing in County Armagh was once again explained by the success of the linen industry, particularly in the northern half of the county, which was as yet a largely domestic affair, although this was about to change. The "linen triangle" had vertices at Lurgan, Dungannon, and Armagh City, although this gradually extended outward as prosperity and demand increased.[61] By 1851, as figure 4.10 shows, the pattern of industrial development had clearly become more intense in the counties of northeast Ulster. This reflected not only the withering of the domestic linen industry as technological and infrastructural advances promoted the development of mills and towns but also the fact that Belfast and the Lagan valley were beginning to see the emergence of heavy engineering and associated industries upon which the economic

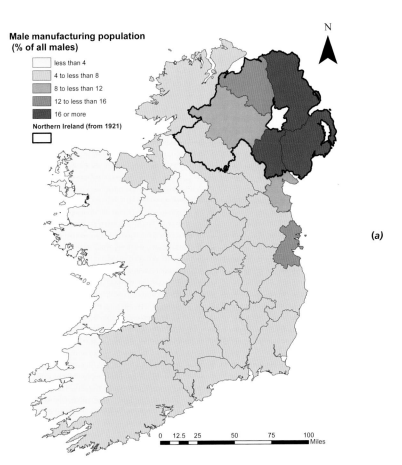

Male manufacturing population (% of all males)

- less than 4
- 4 to less than 8
- 8 to less than 12
- 12 to less than 16
- 16 or more

Northern Ireland (from 1921)

N

(a)

0   12.5   25   50   75   100
Miles

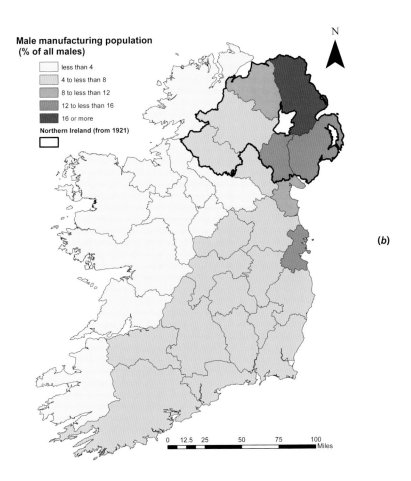

Male manufacturing population (% of all males)

- less than 4
- 4 to less than 8
- 8 to less than 12
- 12 to less than 16
- 16 or more

Northern Ireland (from 1921)

N

(b)

0   12.5   25   50   75   100
Miles

Fig. 4.10. Males employed in the manufacturing sector as a percentage of the entire male population at the county level in (*a*) 1851 and (*b*) 1871.

preeminence of northeast Ulster would be based for the following century.[62] There had been a shipbuilding industry in the city almost from its foundations, but it was in the second half of the nineteenth century that the sector began to expand massively with the development of iron-hulled ships.[63] Dublin stands out in the south of Ireland as the only county with an industrial population of any significance. Dublin's industrial sector was also developing, but on lines very different from those in Ulster. This sector was based largely on food-processing industries and reflected the overwhelming dominance of agriculture and food production not only in the rural economy but in the south's largest towns and cities as well.[64] By 1851 the Irish rail network was still in its infancy, although improvements were beginning to filter through.[65] Industrial developments in Dublin, as the center of an emerging secondary sector based on food, permeated into the hinterland, and the surrounding region also started to develop mills and other processing infrastructures.[66]

Evident also from figure 4.10 is a growing division in terms of industrial employment between the east and west of the county. In 1851 this division could be perhaps explained by the partial impact of improvements in transportation and access to the British market, but by 1871 no such reasons existed. The division between east and west was just as pronounced, and, if anything, the 1871 map indicates that the western part of Ireland was deindustrializing. This had not been a short-term trend but had been developing since the start of the century, and it was the strongest argument that Daniel O'Connell and his supporters had for repeal of the Act of Union. The economies of Britain and much of Ireland had failed to converge in the manner that had been anticipated, and the United Kingdom and the island of Ireland were becoming more and more economically stratified and imbalanced.[67] This is underlined by the performance of the Belfast region, evident from the 1871 map. Industrial employment had become much more concentrated in the county of Antrim by 1871, and that may seem to indicate that similar processes of stratification were at work there. This analysis fails to take account of the powerful multiplier effect that the region was having on the wider economy of Ulster, leading to the development of ancillary tanneries, rope works, iron foundries, and engineering works, not to mention yet more linen mills. The rope works was initially a spin-off reliant on demand from the Belfast shipyards, but it rapidly developed its own markets and soon became the largest in the world.[68] The push-pull factors of Famine hardship and employment opportunity led to a large influx of migrants into the city in the mid-nineteenth century, mainly from across the province, raising the Catholic proportion of the city's population. This area requires further research, but it is usually the case with migration that the remittances sent back by young city workers help to sustain the economies and communities in the rural areas from which those young people departed.

Conclusions

This chapter opened by stating that the Great Famine was an event of unparalleled magnitude, and we have traced the remarkable inconstancy in its spatial imprint in a number of different spheres. In some ways the Famine is a watershed in Irish history; in other ways it is not. In some respects it marked a fundamental change in the country's development; in others it was merely an alteration of a trajectory or an acceleration of trends that were already gathering pace. The previous chapter showed that Ireland was already dividing economically, religiously, and demographically. The Famine, through its partial impact, threw those divisions into sharper relief. The inadequacy of the British response called into question the whole idea of a United Kingdom, while the reluctance of Ulster to subsidize the cost of relief elsewhere in Ireland underlined the lack of fundamental unity already apparent within the island.[69] So Cullen's description of the Famine as more of a regional than a national disaster is accurate in another, albeit unintended sense: Ireland's experience of, and reaction to, the disaster was indicative of an island that was behaving more like a collection of increasingly divergent spatioreligious regions rather than a cohesive entity.

5 Toward Partition, 1860s to 1910s

It is clear that the Great Famine of 1845–51 had a profound effect on Ireland, leaving its mark on a significantly altered and diminished society. It is also clear that the Famine's impact was not uniform across the entire island. The death and dispersal it caused were catastrophic, but the processes it set in train were just part of an ongoing demographic tragedy for Ireland. The extent to which the Famine was a watershed in these events, or simply acted to accelerate preexisting trends, remains controversial. Nevertheless, it is clear that in the second half of the nineteenth century—the post-Famine period—significantly different paths were followed by the northern and southern parts of Ireland. This led to a divide that encompassed economic, political, ethnonational, and religious aspects, and all of these had distinct and interrelated geographies. This mix would explode in the early twentieth century.

Population in the North of Ireland

Mary Daly's book *The Slow Failure* deals with independent Ireland's continued population decline in the twentieth century.[1] The fall in population that she discusses was a continuation of an ongoing trend that had been set in place by the Great Famine and that gathered pace during the latter half of the nineteenth century. Despite the virulence of the demographic hemorrhage, it soon became apparent that the differing spatial impacts of the Famine earlier noted were leading to lasting differences in the event's historical footprint.

Figure 5.1 shows demographic trends throughout the island of Ireland over the period 1821 to 1911 broken down by the spatial units that would become the separate political entities after Partition. The graph illustrates how much more profoundly the area that was to become the Republic of Ireland was affected by population decline subsequent to the Famine than what was to become Northern Ireland. While the population of the six northern counties had stabilized by 1901, south of what was to become the border the population continued to decline for decades. Table 5.1 elucidates these findings a little further. The initial fall between 1841 and 1851, by which point the full immediate effects of the Famine could not even yet be properly enumerated, was much more severe in the south than in the north, with the Republic of Ireland area dropping by 21.7 percent compared to just 12.5 percent for the Northern Ireland area. The disparity in decline continued, with only the 1881 census showing a convergence in the rate of decline north and south of the future border. Afterward the pattern

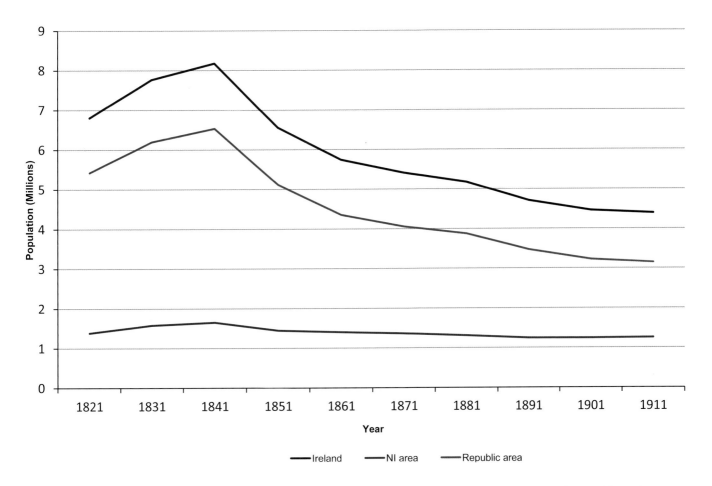

Ireland ——NI area ——Republic area

diverged further, with the south returning toward more rapid decline, while the north headed toward population stability and, finally, growth.

The series of maps in figure 5.2 show the impact of population decline on population density. The maps for 1871 and 1891 are at the barony level. The baronies had been replaced by the rural and urban districts by the turn of the century, as described in chapter 1.[2] Comparing figure 5.2 with figure 4.1 shows that over the period 1861 to 1891 there appears to be a continuation of the pattern of population loss across the rural areas of mid-Connacht, mid-Munster, and mid-Ulster. Even the most densely populated and prosperous area of rural Ireland, the linen triangle of north Armagh, was showing signs of serious population decline in this period. Figure 5.3

Fig. 5.1. Long-term demographic trends throughout the island of Ireland and for the areas that would later become Northern Ireland and the Republic of Ireland.

Table 5.1. Population figures and percentage change in population for the island of Ireland and the Northern Ireland and Republic of Ireland areas

	1821	1831	1841	1851	1861	1871	1881	1891	1901	1911
Ireland*	6,802	7,767	8,175	6,552	5,746	5,412	5,175	4,708	4,459	4,392
RoI area*	5,421	6,193	6,529	5,112	4,350	4,053	3,870	3,469	3,222	3,141
NI area*	1,380	1,574	1,646	1,441	1,396	1,359	1,305	1,239	1,237	1,251
RoI area % change		14.2	5.4	−21.7	−14.9	−6.8	−4.5	−10.4	−7.1	−2.5
NI area % change		14.1	4.6	−12.5	−3.1	−2.7	−4.0	−5.1	−0.2	1.1

* Population in 1,000s.

Fig. 5.2. Population density in Ireland for (a) baronies in 1871, (b) baronies in 1891, and (c) rural and urban districts in 1911.

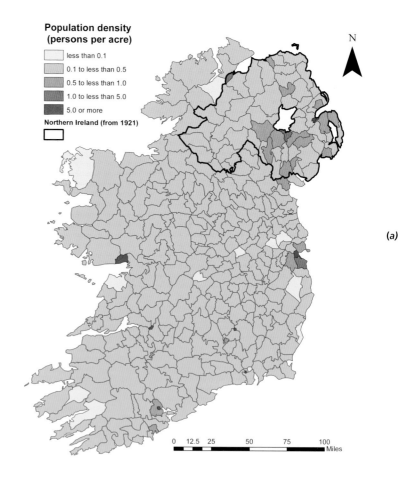

(a)

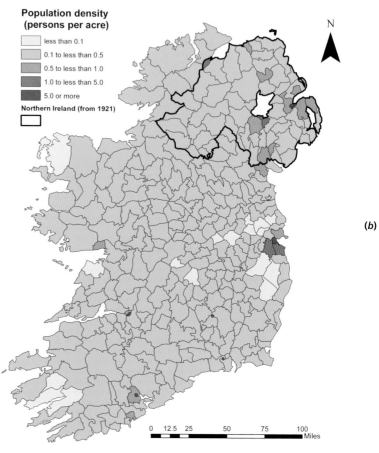

(b)

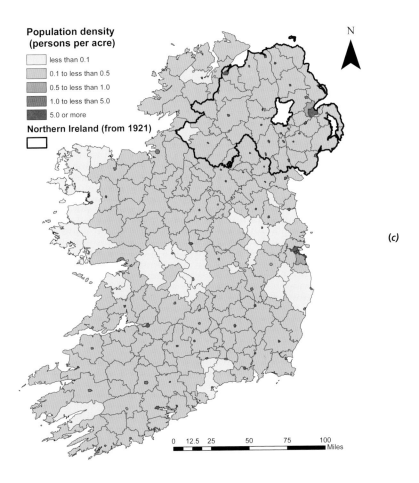

**Population density
(persons per acre)**

less than 0.1

0.1 to less than 0.5

0.5 to less than 1.0

1.0 to less than 5.0

5.0 or more

Northern Ireland (from 1921)

N

(c)

0 12.5 25   50   75   100
                                    Miles

and table 5.2 explore the population trends for this county in a little more detail, as they prove extremely revealing in terms of the broader economic and demographic processes that were shaping this part of the island in the late nineteenth century.

First, it is apparent that despite its history of prosperity based on the domestic linen industry, County Armagh was not immune from the large-scale population loss that had occurred across all of rural Ireland in the post-Famine period. In the half-century between 1861 and 1911, the county's population fell by over 36 percent. However, the transition to census enumeration using rural and urban districts by 1901 enables us to identify the dynamic of urban growth hidden by the picture of population decline. Between 1901 and 1911 the population of the county as a whole fell by 4 percent, while the urban population grew by 7 percent. While the worst effects of the Famine were less keenly felt in north Armagh, the collapse of the domestic linen industry did most profoundly affect this region. So the "push" factor of rural economic hardship and the "pull" factor of increasing urban opportunity conspired with the relative accessibility of the Belfast region to north Armagh to depress the rural population of the area. While the growth of towns like Lurgan and Portadown may appear noteworthy in an overall context of demographic decline, the fact that Belfast's population doubled between 1861 and 1891 and then increased by the same amount again over the next twenty years indicates just how powerful the dynamics of urbanization were.[3]

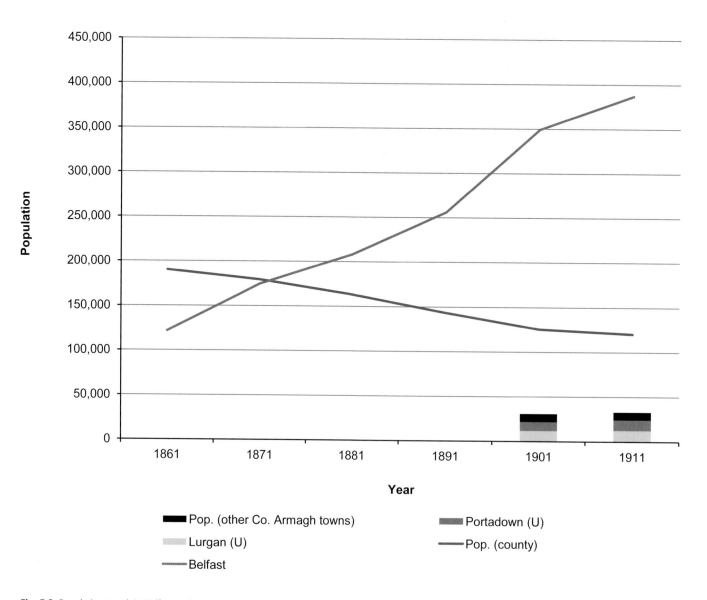

Fig. 5.3. Population trends in Belfast and County Armagh between 1861 and 1911.

Table 5.2. Population trends in Belfast and County Armagh between 1861 and 1911 in tabular form

	1861	1871	1881	1891	1901	1911
Pop. (county)	190,086	179,260	163,177	143,289	125,392	120,291
Pop. (urban)					30,889	33,045
Lurgan (Urban)					11,782	12,553
Portadown (Urban)					10,092	11,782
Pop. (other County Armagh towns)					9,015	8,710
Belfast	121,602	174,412	208,122	255,950	349,180	386,947

Industry in the North of Ireland

It is impossible to separate demographic processes from major economic developments in Ireland at this time. Figure 5.4 shows male employment in manufacturing in 1891 and 1911, a time period that follows from figure 4.10, which shows the same data in 1851 and 1871. These data show how Ireland's manufacturing industries were increasingly concentrated in Antrim, more specifically Belfast, while even Armagh declined. This pattern is closely

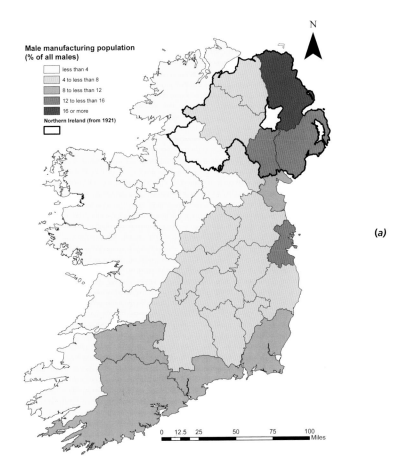

**Male manufacturing population
(% of all males)**

- less than 4
- 4 to less than 8
- 8 to less than 12
- 12 to less than 16
- 16 or more

Northern Ireland (from 1921)

N

(a)

0 12.5 25   50   75   100
Miles

Fig. 5.4. Males employed in the manufacturing sector as a percentage of the male population at the county level in (*a*) 1891 and (*b*) 1911. The patterns for 1851 and 1871 are shown in figure 4.7.

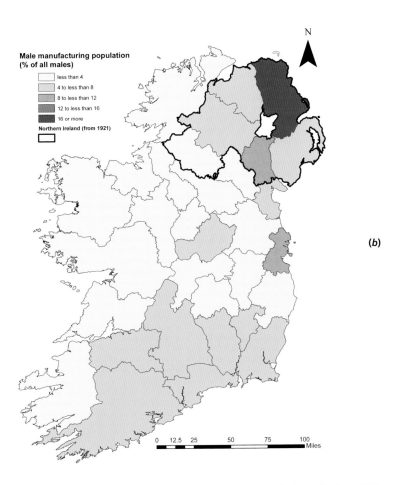

**Male manufacturing population
(% of all males)**

- less than 4
- 4 to less than 8
- 8 to less than 12
- 12 to less than 16
- 16 or more

Northern Ireland (from 1921)

N

(b)

0 12.5 25   50   75   100
Miles

related to the population contractions and expansions noted above, which in Ulster were heavily reflective of the restructuring taking place within the linen sector at the time as it transformed into an increasingly mechanized and geographically concentrated industry in the late nineteenth century.

The picture of population change may suggest that the linen industry was in crisis, and periodically this was indeed the case, but during the course of the late Victorian era Ulster managed to position itself at the very center of world textile production. In 1850 just over 21,000 people were employed in the cotton and linen sector in 69 factories throughout Ireland, with the vast majority being in northeast Ulster, focused heavily on the Lagan valley. By 1890 there were some 162 plants employing close to 65,000 people.[4] Linen was thus a rapidly growing industry. Infrastructural improvements to the road and rail networks had greatly improved access to markets and also allowed the industry to disperse somewhat. This helps to explain the growth in regional urban centers like Lurgan and Portadown and also in much smaller settlements such as Bessbrook and Gilford in County Down whose sole economic raison d'être was linen manufacturing, specifically, the making of handkerchiefs.[5] There was a degree of serendipity in the advance of the linen industry in Ulster, as the American Civil War in the early 1860s crippled the cotton industry in Britain as imports of the raw material collapsed. This created a huge gap in the textile market that Ulster was able to fill with linen.[6]

While Ulster's industrial achievements at this time were considerable, it was only in linen that it truly gained preeminence, not only in a British but in a global context.[7] In addition to linen, in the second half of the nineteenth century Belfast became synonymous with heavy engineering, particularly shipbuilding. Many Irish port towns had a tradition of shipbuilding, but it was only Belfast, partially benefiting from its proximity to the port of Liverpool, that developed a profitable trade based on the construction of ocean liners.[8] Innovative design and engineering methods developed by Harland and Wolff Heavy Industries Ltd. enabled the company to win the contract in 1869 to supply ships to the White Star Line, of which the *Titanic* was to become the most famous.[9]

Shipbuilding was just one facet of Belfast's transformation into a major industrial center in this period. The requirements of the increasingly mechanized linen and shipbuilding industries spawned a powerful engineering sector dominated by names such as James Mackie and Sons and MacAdam and Brothers' Soho Foundry.[10] What is critical about the industrial character of northeast Ulster in the late nineteenth century is the extent to which it faced outward and how its fortunes were so heavily tied up with the British imperial project across the globe. Processing tobacco, manufacturing irrigation pumps for the Nile delta, or producing tea-drying systems for India tied Belfast directly into the colonial enterprise and allied the interests of Ulster to those of Britain in ways that were both economic and ideological.[11] Ulster's growth was explicitly tied up with the rapid expansion of Britain's largest trading port, Liverpool.[12] The net register tonnage passing through this city, which served the entire empire, from 1861 to 1911 exploded from 5 million to 17.6 million.[13] Shipbuilding became emblematic of Belfast's success because it symbolized most clearly the link between the city and the wider empire.

Employment in engineering, shipbuilding, and other heavy industries was completely dominated by men, while the workforce in the linen industry was predominantly female. The industries therefore complemented each other and facilitated an improvement in the living conditions and wealth of the growing Belfast working class.[14] Analysis of patterns of female employment in manufacturing reveals even more starkly the economic differences emerging in the northern and southern parts of the island. Figure 5.5 shows female employment in manufacturing as a percentage of the female population between 1871 and 1911. The maps show a remarkably high degree of economic activity amongst women in Ulster compared to the rest of the island, and this activity becomes increasingly concentrated in Antrim and Armagh. Once again, this reflects the preeminence of the linen industry in the north and the fact that its success was built in no small part on the low wages paid to women and children working in the mills. These maps measure female engagement in manufacturing against the entire female population because it is difficult during this period to establish a consistent measure of working age due to incremental reforms designed to protect workers during the late Victorian period. The minimum working age in the mills was just eight years old prior to 1874. By 1901 it was only twelve years old.[15]

While the gender structure of industrial employment in Belfast may have worked to the advantage of the urban working class, the idea that the city was a place of social harmony in the late nineteenth century could not be further removed from reality. As we shall see, rather than acting to ameliorate sectarian tensions between Catholics and Protestants, Belfast's industrial might only served to underline the disparities between them.

Outside of the major towns and cities in the south, employment opportunities for women were largely characterized by unpaid labor on the family farm. Even by 1912 only 26 percent of men and women working in agriculture were classified as wage earners.[16]

Population and the Economy in the Rest of Ireland: The "Slow Failure"

As the northeast became increasingly dependent on external trade, its economy became disengaged from that of the rest of the island. Figure 5.4, together with figure 4.10, shows that during the period from 1871 to 1911 the percentage of males employed in manufacturing was declining. This was not simply the case in the west, where the sector had always been insignificant, but was progressively so in the east of the island as well. Manufacturing employment in the south of Ireland had historically been heavily concentrated in Dublin City. However, even this important center shows a substantial decline in the number of men working in the production of goods during the same period, down from 12.9 percent in 1871 to 9.6 percent by 1911. By far the largest industrial concern in Dublin was the Guinness brewery, which had become the largest in Europe by the turn of the century. Guinness had come to dominate beer production in Ireland largely at the cost of other smaller producers.[17] The brewing tradition in Cork managed to withstand the commercial onslaught of the Dublin giant, although in 1900 the combined sales of Murphy's and Beamish & Crawford

Fig. 5.5. Females employed in the manufacturing sector as a percentage of the female population at the county level in (a) 1871, (b) 1891, and (c) 1911.

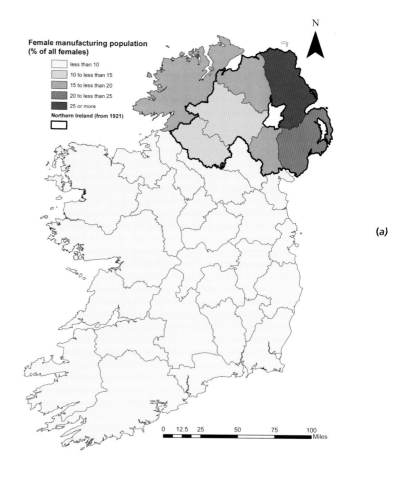

Female manufacturing population
(% of all females)

less than 10
10 to less than 15
15 to less than 20
20 to less than 25
25 or more

Northern Ireland (from 1921)

(a)

0 12.5 25 50 75 100
Miles

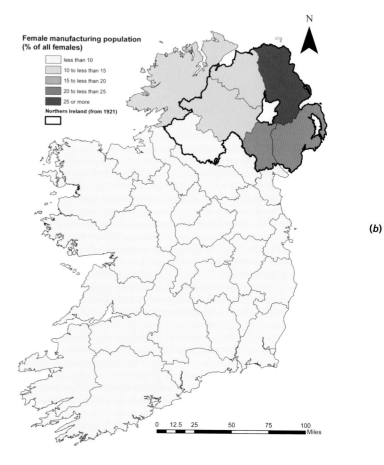

Female manufacturing population
(% of all females)

less than 10
10 to less than 15
15 to less than 20
20 to less than 25
25 or more

Northern Ireland (from 1921)

(b)

0 12.5 25 50 75 100
Miles

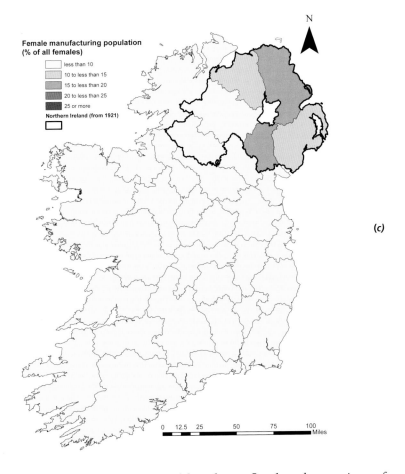

N

(c)

0   12.5   25   50   75   100
                                    Miles

was only one-eighth of Guinness and largely confined to the province of
Munster.[18] The fact that brewing and, to a slightly lesser extent, distilling
were the most successful industries in the south of Ireland was a source
of pain rather than pride to many. Alcohol consumption was historically
extremely high in Ireland, although it was falling.[19] In part this decrease
was aided by the well-established temperance movement, which had allied
itself with some success to wider political and religious objectives.[20]

Why did industrial production in the south of Ireland appear to stagnate in the post-Famine period? Academics have long been exercised by
this aspect of modern Irish history but have struggled to offer convincing
explanations for the differing patterns of economic development on the
island from the mid-nineteenth century. They have been more successful
in discounting reasons for the uneven patterns of industrialization. For example, Roy Foster states that an absence of indigenous raw materials was
not an issue, as Ulster was not hampered by an almost complete reliance
on British coal and iron ore. Nor was there a dearth of sufficient capital for
investment. He alludes to a potential cause when he argues that while individual capitalist concerns in the south often flourished, the profits were
often ploughed back into "land and the appurtenances of gentility."[21] The
Anglo-Irish professional and entrepreneurial class may "have had an influence out of all proportion to its exiguous numbers," but it failed to provide
the sort of commercial leadership that its coreligionists supplied in abun-
dance in Ulster, while those Catholics who aspired to this social echelon
appeared more concerned with acquiring the baubles of success rather than
developing their enterprises.[22]

It should also be remembered that the Anglo-Irish elite was a largely rural caste with a reputation in Britain for idleness rather than innovation. The Ulster poet Louis MacNeice contemptuously dismissed the big house culture of the Anglo-Irish as containing "nothing but an obsolete bravado, an insidious bonhomie and a way with horses."[23] He was writing in the twentieth century, but the remark was just as accurate of late nineteenth-century Ireland given the anachronistic lifestyle of the Anglo-Irish. Joseph Lee dismisses the notion that the cause of Ireland's failure to industrialize lay in part in the inefficiencies of the food production and supply system, which meant that not enough surplus food was being produced to sustain a large, industrialized, and, by natural extension, urbanized workforce. The Irish agricultural sector had moved from being primarily subsistence driven in the pre-Famine period to a market-based model in the post-Famine period as Irish farmers responded to changed circumstances, as evidenced by the move from tillage to pasture across much of the country. Yet despite this responsiveness, Irish farmers' hands were tied by the system of land tenure, and this was the primary determinant of their ability, and perhaps motivation, to modernize.[24]

The particular quality of entrepreneurship in the northeast has often been cited as a key factor in the region's development, but Lee rejects the idea that this is testament to Ulster's innate "Protestant ethic," as most of the key individuals concerned, such as James Harland, Gustav Wolff, George Smith, and William Pirrie, were born beyond Ireland's shores.[25] Perhaps the power of personality in determining the northeast's economic fortunes should not be underestimated. It may well be the case that a lack of ambition stymied the industrial development of cities like Dublin. Taking shipbuilding as an example, there would have been no industry to speak of had Belfast's city fathers not had the initiative to drain low-lying mudflats and protect them with dykes, creating flat sloblands for development around Belfast Lough and clear deep channels for navigation.[26] By the time of the Famine, the maritime improvements on which the city's success was built in the second part of the century were already well advanced. Dublin in 1900, on the other hand, had not even the facilities to repair its own merchant fleet, much less the infrastructure to promote the sort of heavy engineering that would employ large numbers of workers.[27]

By 1911 the north of Ireland was continuing to expand based on the core industries of linen, engineering, and shipbuilding, while, with the notable exception of food and drink, the south of the country was going in the opposite direction.

Emigration

Emigration continued to play a powerful role in Ireland during the period 1861 to 1911. By the late nineteenth century it had become the default response to economic hardship, and the extent to which it had by then become ingrained in the Irish psyche cannot be overstated.[28]

One of the interesting features of the maps in figure 5.6, showing emigration rates in 1871, is the relatively high rate from the northeast of the island. This surprising finding is difficult to explain and may simply be a

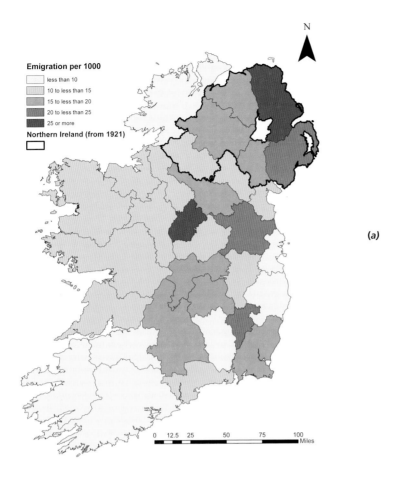

Emigration per 1000

	less than 10
	10 to less than 15
	15 to less than 20
	20 to less than 25
	25 or more

Northern Ireland (from 1921)

N

0 12.5 25   50   75   100
Miles

(a)

Fig. 5.6. Emigration rates per 1,000 in 1871 for (a) males and (b) females.

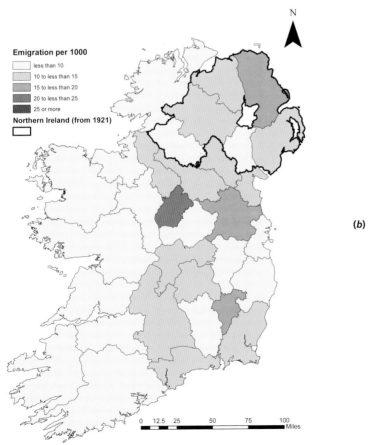

Emigration per 1000

	less than 10
	10 to less than 15
	15 to less than 20
	20 to less than 25
	25 or more

Northern Ireland (from 1921)

N

0 12.5 25   50   75   100
Miles

(b)

Toward Partition, 1860s to 1910s

result of the fact that the maps provide a snapshot of emigration patterns in one single year and thus may be skewed by short-term processes. David Fitzpatrick has found that Antrim was consistently ranked lowest in terms of the cohort depletion measure of emigration in the period 1871 to 1911.[29] This suggests that the 1871 pattern was the exception rather than the rule. There were two contradictory dynamics within Belfast's economy at this time: the massive growth of shipbuilding, as Harland and Wolff was building the world's first modern ocean liners for the White Star Line, on the one hand, while at the same time there was a depression in the linen industry after the boom of the late 1860s brought about by the American Civil War.[30] The fact that emigration seems to have disproportionately affected the male population does not seem to fit this pattern, as female employment would be expected to have suffered the most from these trends. Despite this, the Belfast region still offered women a wealth of employment opportunities that existed nowhere else on the island, which may help to explain the low female emigration rates as compared to males.[31]

From 1881 the pattern of emigration settled into a more predictable mode as it increased in the western parts of Ulster and Ireland more generally. Figure 5.7 illustrates the data for 1891. The abiding pattern of emigration from Ireland was from counties where agriculture was the primary source of employment and where the agricultural landscape was characterized by small farms of low value.[32] The highest rates of emigration were consistently found in the west of the country. Particularly striking are the scale and uniformity of female emigration in 1891. A number of explanations for this pattern can be advanced. First, and perhaps most obviously, the financial benefits of leaving Ireland for North America, Britain, or elsewhere were seen to be far greater than staying put, given that, as already established, many women (and men) were not even operating within the wage economy in rural Ireland.[33] Another explanation may lie in the fact that women in more remote parts of the country had poorer physical and social access to the employment networks that operated around the main cities in the east and as a result naturally militated to Britain or, more likely in this period, westward to the United States.[34] A third reason may lie in the hardship of everyday rural life, and this explanation applies equally to both male and female emigration. In 1936 R. C. Geary argued that people left rural Ireland despite innovations in agriculture simply because they wanted to rather than because of basic economic necessity.[35] He also argued that emigration by this stage had become an accepted cultural and social norm, quoting a writer who said, "Connemara children are born with their faces towards the West," meaning that they were already looking forward to a future overseas in North America.[36] It has already been stated that Irish agriculture was moving from labor-intensive tillage to more profitable livestock farming. This shift led to a drop in the demand for agricultural labor, and thus these are the regions where the rural population fell most sharply.[37] This notion that economic imperatives alone did not form the only reason for migration is borne out by the evidence of one agricultural laborer in 1894 who told an enquirer, "I don't like the work on the land. It is very laborious and doesn't lead to anything."[38] For many it certainly did not lead to anything, as changes in inheritance structures after the Famine meant

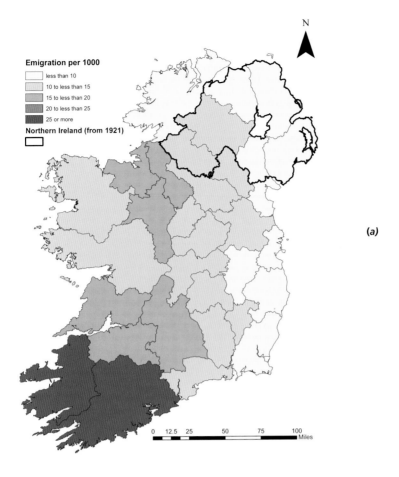

Fig. 5.7. Emigration rates per 1,000 in 1891 for (*a*) males and (*b*) females.

(*a*)

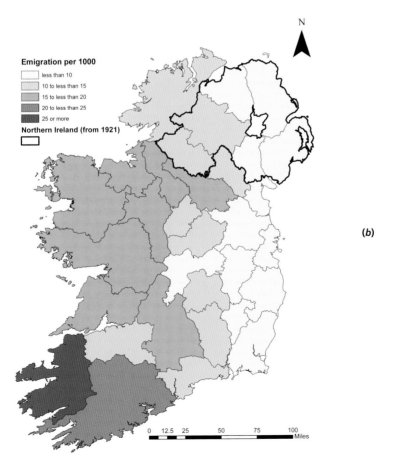

(*b*)

that fewer people were acquiring land and therefore had even less reason to remain in the country. Yet the decline in those inheriting did not lead to the consolidation of farms into more substantial holdings. The farm owners were not emigrating, but their brothers and sisters were.[39]

Geary's analysis touched a raw nerve in early postindependence Ireland, but the truth of the matter probably lies in a combination of all these explanations, with the decision to emigrate being dictated not only by economic necessity and the limitation of employment opportunities in rural Ireland but by some degree of personal choice.[40] This is not to trivialize the circumstances of the rural peasantry and the extent to which their very existence was still tied to the success of the potato crop, even after all the lessons of the Great Famine. The spike in emigration in the west evident in the maps for 1891 came against the backdrop of a serious crop failure the previous year, and it is little wonder that so many opted to leave, given the memories of the years between 1845 and 1851.[41] Not only that, but the fact was that little had changed in the economic circumstances of the rural poor since the Famine.[42] Had they seen a real improvement over the intervening forty years, then perhaps more would have been persuaded to stay. It is this fact that underlines the statistical analysis of Hatton and Williamson, who in surveying emigration trends from 1850 to 1914 found a stronger correlation with much broader demographic and standards of living variables than with variables simply associated with agriculture.[43]

Religious Change

What, then, were the spatioreligious implications of this socioeconomic divergence? Figure 5.8 shows the percentage of Catholic population at the county level, the most detailed geography available in digital form in this period. Given the complete dominance of only two religions, the distribution of Protestants can be taken to be the inverse of the pattern for Catholics. At first glance the time series suggests a remarkable stability in religious geographies. The long-term patterns of Protestant plantation settlement and English influence are still clearly evident in Ulster and in the Pale and there is little evidence of change over the forty years shown in the maps.

The maps, however, mask some of the more complex trends within spatioreligious change. As the previous chapter identified, Ireland's population fell by 27 percent in the twenty-seven years from 1834 to 1861, and this loss was heaviest amongst the Catholic population, which fell by 30 percent compared to 16 percent for Protestants. Over the next fifty years—from 1861 to 1911—the population fell by a further 24 percent, and again the Catholic population fell more rapidly, with a 28 percent loss compared to a 15 percent loss of Protestants. Thus, while the speed of population loss fell, the national-level trends for the total population and within the main religions very much continued. This led to the Catholic majority within Ireland falling further, from 81 percent in 1834 to 77 percent in 1861 and then to 74 percent in 1911. The Church of Ireland's share of the population rose from 10.7 percent in 1834 to 11.8 percent in 1861 and to 13.1 percent by 1911, while the Presbyterian Church's share rose from 8.1 percent to 9.1 percent to 10 percent in 1911. These increases are, however, caused by a lower level of loss of Protestants compared to Catholics rather than a numerical increase in the Protestant population.

Figure 5.9 shows that the distribution of the changes in the 1861 to 1911 period was far from uniform and in many ways followed and exacerbated the changes that took place over the Famine period. Overall population loss continued to be widespread, but, if anything, areas of the northern midlands, stretching as far north as Tyrone, had the greatest losses. Only two counties, Antrim and Dublin, show any growth, but in both cases this is dramatic, at 57 percent and 33 percent, respectively. This, of course, reflects the rapid urbanization of Belfast and Dublin. The patterns of change of the three main denominations largely follow this trend, although the Church of Ireland shows more pronounced decline over much of Ireland and more dramatic increases in Antrim, whose Church of Ireland population grew 114 percent compared to growths of 31 percent and 29 percent, respectively, for Catholics and Presbyterians.

The overall effect of these changes was to increase the spatioreligious polarization within Ireland. The Catholic proportion of the population of the six counties that were to become Northern Ireland fell from 40.9 percent in 1861 to 34.4 percent in 1911. Much of this was because of change in Antrim and Down. In the remaining four counties—Londonderry, Armagh, Fermanagh, and Tyrone—the Catholic proportion only changed from 51.6 percent to exactly 50 percent. In the counties of Dublin and the Pale—Kildare, Wicklow, Carlow, and King's and Queen's Counties—the Catholic proportion of the population remained roughly constant, just falling from 82.6 percent to 81.1 percent. Across the rest of Ireland, effectively what was to become the Republic apart from Dublin and the Pale, the Catholic population again remained roughly constant at 93.9 percent in 1861 and 94.3 percent in 1911. Thus, overall, we see the Protestant proportion of the population growing in northeast Ulster, while over the rest of Ireland the proportions remain roughly the same.

The consequence of this was that measures of segregation also increased in this period. The index of dissimilarity comparing Catholics and Protestants rose from 53.6 percent for counties in 1861 to 57.8 in 1911.[44] The isolation index for Catholics stayed approximately constant at 84.5 percent and 83.4 percent at the two dates, but over the same period the expected isolation index for Catholics fell more sharply, from 77.9 percent to 73.8 percent, suggesting that while the proportion of Catholics fell, this did not make them any more integrated with the Protestant population. Meanwhile, the isolation index for Protestants rose from 42.7 percent to 46.4 percent against an expected index of 20.8 to 23.2, suggesting that Protestants were becoming more concentrated in certain areas than their overall increase in proportion of the population would suggest.

Beyond these simple statistical measures this was a period of intense polarization in Irish society. Divisions between Catholics and Protestants became increasingly merged with the powerful competing ideologies of nationalism and unionism. Thus, ideologies forged against a background of social, economic, and religious upheaval would come to the brink of outright conflict by the end of the period.

Fig. 5.8. Catholic populations at the county level in (*a*) 1871, (*b*) 1891, and (*c*) 1911.

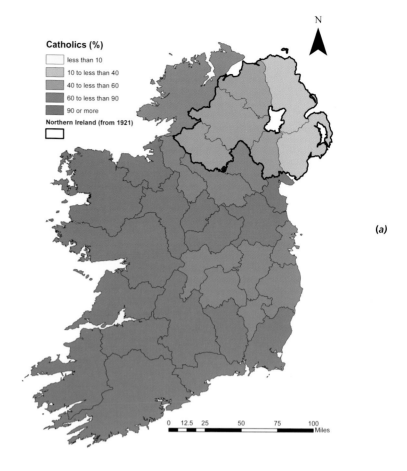

(*a*)

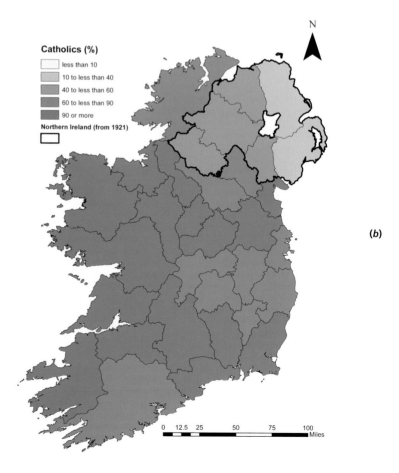

(*b*)

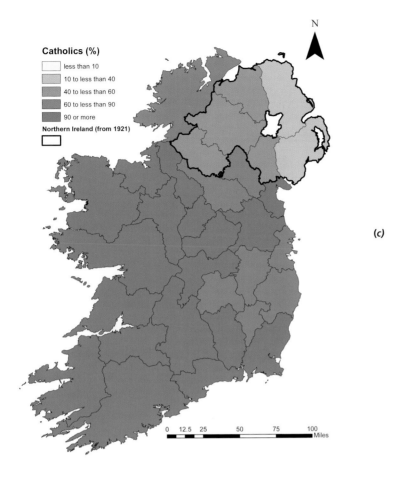

N

(c)

0 12.5 25 50 75 100 Miles

Religion, Politics, and Identity: A Gathering Storm

It is crucial to understand just how much religious tensions in late nineteenth-century Ireland were cross-cut by class and economic issues. It can even be argued that it is more appropriate to invert this and say that economic disputes often exacerbated existing religious tensions between Catholics and Protestants. This was the case with the primary economic issue of Victorian Ireland, that of land reform. The Famine had put land reform firmly on the agenda, but it was not a nettle that successive governments were willing to grasp until they were compelled to by the rise of the Land League. In 1879, against the background of a crop failure that threatened to turn into another Famine, the Land League was symbolically established in Ireland's poorest county, Mayo. The main personality behind the movement was Michael Davitt, a brilliant organizer whose personal experience of both rural and urban hardship had engendered a profound empathy with the plight of the agrarian and industrial worker.[45] He was born to paupers at the height of the Famine in Mayo and brought to England after eviction, losing his right arm in a cotton mill in Lancashire at the age of twelve.[46] The leader of the organization was Charles Stewart Parnell, soon to become the political leader of Ireland's Catholics.[47] The choice of Parnell was significant in two respects. First, as a landlord of Anglo-Irish stock he typified the group that had most to lose from the redistribution of land. Second, by working himself into a position as de facto leader of the Home Rule Party he signaled the emergence of a broader nationalist

Fig. 5.9. Population change at the county level from 1861 to 1911: (a) total population, (b) Catholics, (c) Church of Ireland, and (d) Presbyterians. Note that counties with fewer than 1,000 Presbyterians in 1861 have not been shaded.

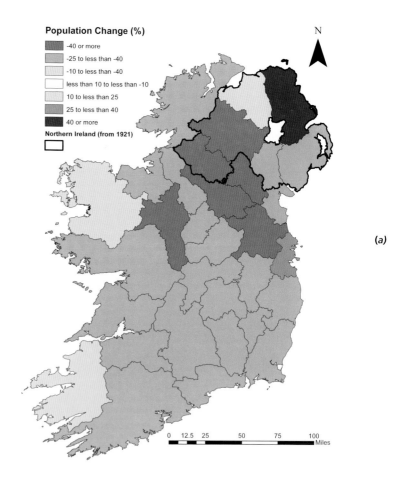

(a)

Population Change (%)

- -40 or more
- -25 to less than -40
- -10 to less than -40
- less than 10 to less than -10
- 10 to less than 25
- 25 to less than 40
- 40 or more

Northern Ireland (from 1921)

N

0 12.5 25 50 75 100
Miles

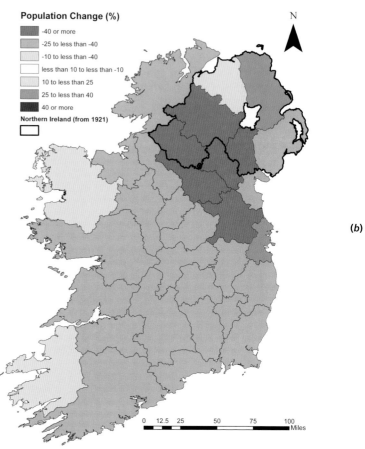

(b)

Population Change (%)

- -40 or more
- -25 to less than -40
- -10 to less than -40
- less than 10 to less than -10
- 10 to less than 25
- 25 to less than 40
- 40 or more

Northern Ireland (from 1921)

N

0 12.5 25 50 75 100
Miles

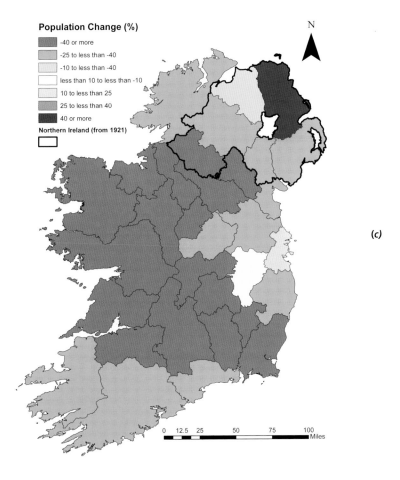

Population Change (%)

- -40 or more
- -25 to less than -40
- -10 to less than -40
- less than 10 to less than -10
- 10 to less than 25
- 25 to less than 40
- 40 or more

Northern Ireland (from 1921)

N

(c)

0   12.5   25   50   75   100
                                      Miles

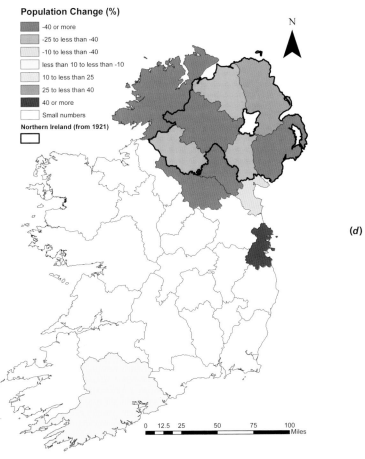

Population Change (%)

- -40 or more
- -25 to less than -40
- -10 to less than -40
- less than 10 to less than -10
- 10 to less than 25
- 25 to less than 40
- 40 or more
- Small numbers

Northern Ireland (from 1921)

N

(d)

0   12.5   25   50   75   100
                                      Miles

alliance.[48] Nevertheless, at this stage the injustices of landownership in Ireland were enough to reconcile Catholics and Presbyterians in their shared grievance at the privileged position of Ireland's Episcopalian Anglo-Irish gentry.[49] This interdenominational unity would not be an enduring feature of Ireland in this period, but it would reemerge at certain critical points, signaling the unrealized potential for a realignment of Irish political culture along class rather than religious lines.

The land issue was effectively removed from Irish politics by the Land Purchase Acts of 1891 and 1896, which provided tenants with the ability to purchase their holdings with a 100 percent loan from the state, subject to an annuity of 4 percent for forty-nine years.[50] By this stage, however, the political atmosphere had altered as demands for a degree of Irish autonomy increased. This trend realized Parnell's longer-term objective of channeling energy on the land issue into the national question.[51] Home Rule had been a fringe issue for many years, but Parnell's ability to marshal the nationalist sentiment of the people, coupled with a widening of the electoral franchise, had served to bring the question to the front and center of the political stage.[52]

Ulster Unionists were appalled by Home Rule, denouncing the prospect of a Dublin government in Catholic-dominated Ireland as "Rome rule."[53] Once the British prime minister, William Gladstone, was converted to the idea, unionists in Ulster began to organize against it in earnest. Gladstone's 1886 bill was thrown out by Parliament, and split his party, but the result provided little solace for Protestants in Ulster.[54] Home Rule slipped off the agenda after Parnell's death in 1891, and Irish nationalism moved from the political to the cultural sphere with a reawakening of interest in the Irish language, sports, and literature.[55] This became known as the Celtic Revival, and although it was not as direct a threat to Protestants as Home Rule, it did serve to alienate them at a cultural level, as it established an exclusive definition of Irishness as being rural, poor, Irish speaking, and ideally located on the western seaboard, aspects of life in Ireland that embraced very few Protestants. The great irony of all this was that so many of the Revival's main protagonists, such as Lady Gregory, Douglas Hyde, and William Butler Yeats, were members of the Protestant Anglo-Irish ascendancy class.

Despite its failure, the 1886 attempt to pass Home Rule precipitated some of the worst rioting in Belfast's history. The riots lasted throughout the early summer and led to the deaths of thirty-two people.[56] This violence was the latest act in an intensifying tragedy of ethnonational conflict in the city. Previous serious outbreaks had occurred in 1857 and 1864 as well as earlier years, and the manner in which they arose tells us much about the growing interdependency of politics, economics, and religion not just in Belfast but throughout the island. The 1857 riots occurred around 12 July, the final day of the Protestant Orange Order's annual season of marches commemorating King William of Orange's victory over the Catholic James II in 1690. This in itself is unremarkable, as violence and increased tension between Catholics and Protestants have been, and indeed still are, a regular feature of the summer marching season. What made 1857 significant was not the religious hostility but the fact that it betrayed the extent to which the communities in Ulster were dividing. The fiery anti-Catholic sermons of the Reverend Thomas Drew acted to provoke his Protestant congregation.[57] The riots were an expression of an increasing insecurity within Presbyterianism in particular that contrasted with a growing self-confidence among the Catholic population in general. The Catholic Church was strengthening its power, status, and control over its congregation through a huge increase in personnel and the introduction of new practices that served only to reinforce the theological and dogmatic differences between Catholics and Protestants.[58] The Catholic Revival was personified by Cardinal Paul Cullen, who harbored a deep antipathy toward Protestantism.[59] Religion was aggravating the turbulent social conditions in the rapidly growing city, as Protestants felt increasingly threatened by the steady increase in the Catholic population.[60]

The story that sparked the 1886 riots goes to the heart of the nexus between economics, politics, and religion. It was said that a Catholic navvy, or unskilled laborer, told a Protestant colleague that, come Home Rule, the tables would be turned, and his kind would struggle to find a job.[61] Months of riots led to the expulsion of Catholics from the shipyards. That such a petty remark could lead to such a conflagration, the worst of the century, tells much of life in Belfast and in Ireland more generally at the time. It not only speaks of the insecurity of the Protestant working class and the threat they felt from the rising (both literally and metaphorically) Catholic population in their city but also says much about the iniquities that existed within employment structures in the city at the time. Protestants felt they would be economically, politically, and religiously subjugated under a Dublin-led government. Their response was to fight for the maintenance of a status quo that subjugated Catholics both politically and economically. The zero-sum game that was emerging in Ulster meant that one group's concession was another's gain, and that would not stand.[62]

The debate over reform of Belfast Corporation—responsible for local government and most evident in the election of councilors and its mayor— serves to support this assertion and goes to the heart of the nexus between religion, economics, and politics. A bill in Westminster Parliament proposed that the city area be extended and be divided into fifteen new wards. During debates on the bill, the fact that Catholics had no representation on the current Corporation despite making up over a quarter of the city's population drew criticism from across the House of Commons.[63] When the mayor admitted that under new arrangements Catholics could not be guaranteed a seat and that not a single Catholic contractor had ever been employed by the Corporation, his statement highlighted the extent of the systemic discrimination that existed.[64] In the end just two seats in the west of the city represented the Catholic minority.[65] The most interesting feature of this debate, however, was the fact that two of the leading unionist political protagonists were also two of the city's industrial godfathers, William Pirrie, mayor of Belfast, and Gustav Wolff, Unionist MP for East Belfast. There was a clear logic here, and it can be summed up in the Belfast Chamber of Commerce's 1893 letter to Gladstone: "All our progress has been made under the union."[66] The defense of unionist political interests was now inseparable from Protestant economic imperatives. Money and both secular and theological beliefs had ossified to create a permanent ideological division on the Irish political landscape.

Conclusions

While there were no major traumas over this period, it was nevertheless pivotal in the island's history as fault lines in Irish society became more polarized. Rural decline over much of the island led to a huge loss of population that the two major urban centers, Belfast and Dublin, could not offset, contributing to a culture of emigration. At the same time these two major urban centers were also following differing trajectories: Belfast was growing rapidly as an industrial city based on textiles and shipbuilding with an economy that was closely tied into the wider U.K. and British Empire economy. Dublin was, if anything, stagnating, and its industry, such as it was, was based on brewing and other agriculture-related industries that tied it more closely to serving rural Ireland.

This economic polarization was also reflected in religious, political, and social polarization. Belfast and its hinterland in northeast Ulster were already the areas with the largest Protestant populations. The growth of Belfast increased this. The Catholic population of Antrim grew from 101,000 in 1861 to 132,000 in 1911, but this represented a decline from 27.5 percent of the county's population to 22.9 percent. The major growth came from the Church of Ireland, whose proportion of the population grew from 20.2 to 27.6 percent over the same period. Presbyterians also showed proportional decline but remained comfortably the largest of the three religions, comprising 47.7 percent and 39.2 percent in 1861 and 1911, respectively. Social divisions often followed religious lines. Much of the industry was owned by Protestants, and discrimination against Catholics in both employment and politics was becoming common. As Home Rule became an increasingly popular political issue, the economic and historical links between this part of Ireland and Britain meant that Protestants were strongly in favor of maintaining the union. Elsewhere in Ireland a lack of manufacturing and an uncompetitive agricultural sector led to population decline and emigration. Land Reform and the Celtic Revival made many of these areas look increasingly to their Gaelic identity, which was closely linked to religion, as the areas in which these took hold were overwhelmingly Catholic. As a consequence, there was a convergence of economic, cultural, religious, political, historical, and geographical forces that increasingly divided Irish-facing Catholics from British-facing Protestants.

Partition and Civil War, 1911 to 1926

6

By the beginning of the twentieth century, division had emerged as the primary motif of Irish society. There were many reasons, both economic and social, for this, but their impact was to divide Catholic from Protestant both psychologically and geographically. The last all-Ireland census occurred in 1911, as Partition was to follow in 1921. The next census took place in both parts of the newly divided island in 1926. From a census perspective, however, 1911 can be regarded as a new beginning, as, despite the fact that since then there have been two separate censuses, sometimes taken in different years, these enumerations provide a number of advantages over those that preceded 1911. The main advantage is that more spatial detail is provided on the geographies of religion that, along with many other variables, are reported at the urban and rural district levels. As described in chapter 1, this provides more districts and also shows the difference between urban areas and their rural counterparts.

Between 1911 and 1926, therefore, we have more geographical detail than was previously available to explore what was the second period of major trauma in modern Irish history. Partition was the culmination of the increasingly incompatible geographies brought about by economic developments and social changes particularly associated with identity and religion. These polarized the political arena as Irish nationalism became increasingly irreconcilable with unionism. The consequences had a strongly sectarian character: the desire for a sustainable Protestant majority was a major factor behind the shape of the newly formed Northern Ireland, which was to remain part of the United Kingdom, while a newly independent Free State set off on its own path.

Population Divergence: Standing Alone

The period between the censuses of 1911 and 1926 was the last in which the population of the island of Ireland as a whole would decline, marking the end of a trend that had lasted for half a century.[1] Thereafter, the population of the island would increase, albeit slowly, throughout the rest of the first half of the twentieth century. Within this period, however, there were different population trajectories between the newly created Northern Ireland and the Free State. These trajectories had been developing since at least 1891, but Partition made them plain for all to see. At the start of the twentieth century the south of Ireland's demographic crisis was already a source of grave concern, but, at least for the time being, Irish nationalists could lay the blame for that crisis at Westminster's door. They argued that the south's failure to industrialize was a result of mismanagement on the part of the British government and that the continued population hemorrhage was a direct consequence of a policy of inaction.[2] The news in 1911 that, as a consequence of its industrial expansion, Belfast had overtaken Dublin as the largest city in Ireland only served to reinforce this belief.[3] If the growth rates of the northern and southern capitals were to be metaphors for the economic well-being of the wider areas, then the figures provided a depressing outlook for Dublin and its hinterland. Figure 6.1 shows the population growth in the two cities between 1861 and 1926. Belfast grew phenomenally in the late nineteenth century, while Dublin stagnated or even declined. From 1891 population growth did begin to pick up in Dublin, but this was due in no small measure to an ongoing process of generously redrawing its boundaries.[4] Thus, by 1911 Dublin had been eclipsed economically and demographically by its northern rival and was left trading on the bygone glories of the Georgian era as its plaintive claims to being second city of the British Empire sounded increasingly hollow.[5]

The maps in figure 6.2 show the population densities of rural and urban districts across the island in 1911 and 1926. There is little discernible difference between the two dates according to these maps, and it is only by comparing the populations directly at district level for each date that we can gauge population trends during this period at the local scale. Figure 6.3

Fig. 6.1. Population change in Belfast and Dublin, 1861 to 1926. * The figures for 1861–91 are combined totals for Belfast (County Antrim) and Belfast (County Down) baronies; the figures for 1901 and 1911 are for the Borough of Belfast. † The figures for 1861–91 are for Dublin City barony; the figures for 1901 and 1911 are for the Borough of Dublin.

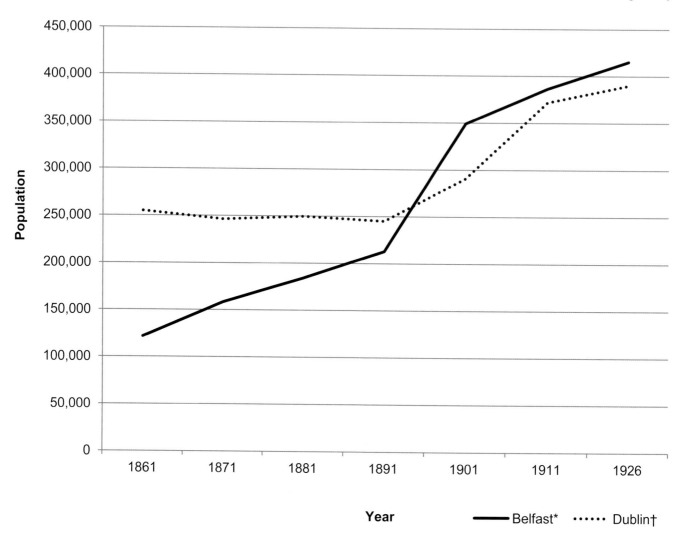

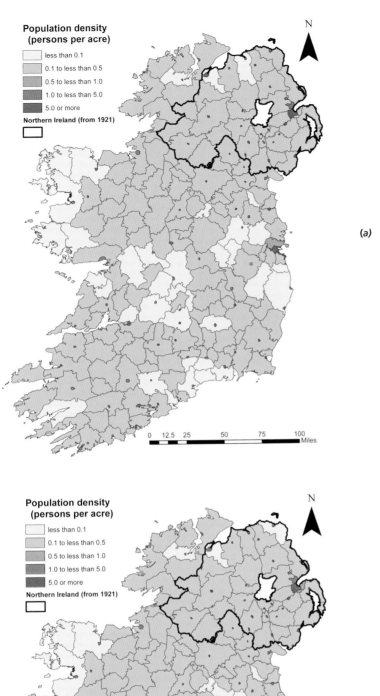

Population density
(persons per acre)

less than 0.1
0.1 to less than 0.5
0.5 to less than 1.0
1.0 to less than 5.0
5.0 or more

Northern Ireland (from 1921)

0 12.5 25 50 75 100
Miles

(a)

Fig. 6.2. Population density of Irish rural and urban districts in (*a*) 1911 and (*b*) 1926.

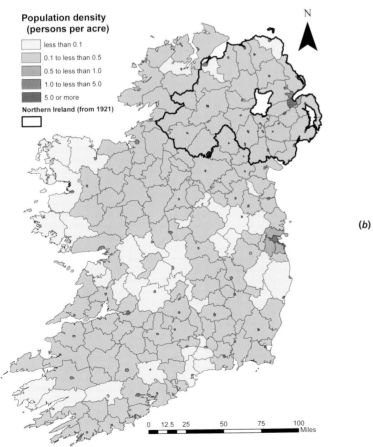

Population density
(persons per acre)

less than 0.1
0.1 to less than 0.5
0.5 to less than 1.0
1.0 to less than 5.0
5.0 or more

Northern Ireland (from 1921)

0 12.5 25 50 75 100
Miles

(b)

Fig. 6.3. Percentage of population change at the rural and urban district levels between 1911 and 1926.

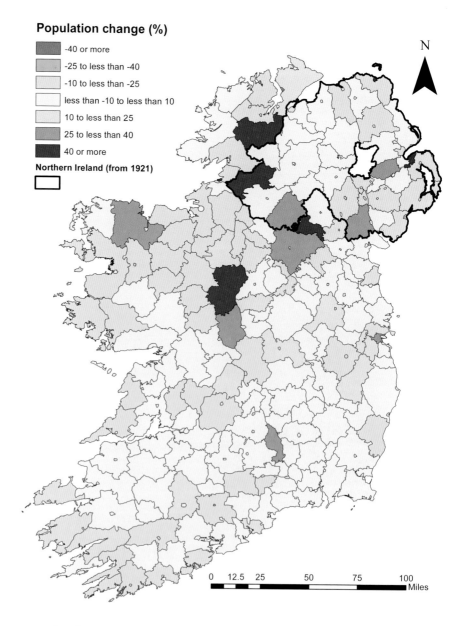

Population change (%)

- -40 or more
- -25 to less than -40
- -10 to less than -25
- less than -10 to less than 10
- 10 to less than 25
- 25 to less than 40
- 40 or more

Northern Ireland (from 1921)

N

0   12.5   25   50   75   100
                                    Miles

provides this information. It shows that, in comparison to the radical falls experienced in earlier periods, fairly broad tracts of the island remained reasonably stable in population terms in the years between 1911 and 1926. The western coastal regions continued to experience population decline, with the largest contiguous area of decline in the northwest, centered on County Sligo. The map reflects the growth of the population around Dublin and Belfast, but other urban centers such as Cork and Limerick did not grow significantly during the period. The map presents a slightly misleading picture, as it suggests that there were pockets of substantial rural growth surrounded by decline, for example, in County Roscommon in the midwest of Ireland. This was not the case. Roscommon saw its population drop by over 11 percent in the fifteen-year period, from 93,856 to 83,556. The supposed growth is actually the product of districts being aggregated together between the two dates. For example, Roscommon rural district had come to incorporate Strokestown rural district by 1926 and therefore saw a significant percentage rise in the population. Overall the cumulative population of the two separate districts had actually fallen sharply. Such

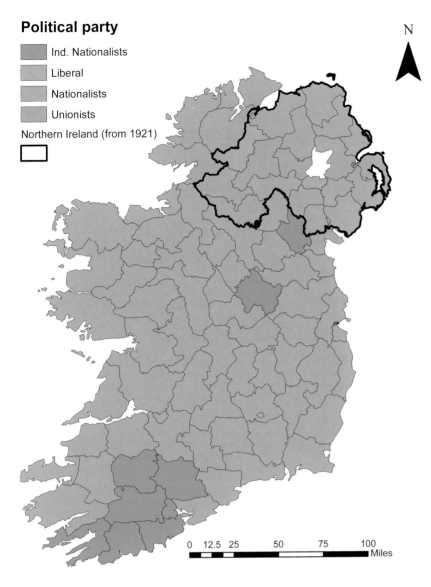

Political party

- Ind. Nationalists
- Liberal
- Nationalists
- Unionists

Northern Ireland (from 1921)

N

0 12.5 25   50   75   100
                              Miles

Fig. 6.4. Results of the December 1910 Westminster elections. Cork City and the University of Dublin were both two-seat constituencies in 1910. Nationalists won both seats in Cork City, Unionists both seats in the University of Dublin.

Party	Seats
Independent Nationalist	9
Liberal	1
Nationalist	73
Unionist	19

aggregations of units are in themselves a comment on demographic trends, as they typically reflect a fall in the population, whereas the subdivision of administrative spaces usually corresponds to population increase. In every case where the map seems to show an increase in the population in rural districts isolated from the major cities, this is actually a product of boundary change reflecting continuing population decline.

Political Divergence: "A Terrible Beauty Is Born"[6]

Ireland changed more during the short period between 1911 and 1926 than at any other time during the span covered by this book. This section provides the political context to the developments that this chapter identifies.

Figure 6.4 shows the results of the December 1910 general election. As with most other maps in this book, the border is shown, even though at this stage few people would have thought that Ireland would be partitioned only around a decade later.[7] Despite this, it is clear that the country was already divided geographically and politically. Within the six counties that were to make up Northern Ireland, the Unionists won seventeen of the twenty-five seats, compared to the Nationalists' seven. Across the remainder of the island, the Nationalists or Independent Nationalists won every seat apart from the two University of Dublin seats that went to the Unionists. Thus politically the island was already divided into a nationalist south that returned candidates supporting Home Rule and a unionist north whose dominant political culture was dedicated to maintaining the link with Britain in its fullest form. For the vast majority of nationalists, agitation for Home Rule was not a halfway house to full independence, for which there was little appetite. Instead it was merely the recognition that Ireland should be allowed to take its rightful place alongside the other member nations of the British Commonwealth.[8] There was no contradiction in the idea of a measure of self-determination under the Crown.

While it was far from obvious in 1911 that the island would soon be divided into two states just ten years later, over the course of the intervening decade it became apparent that few other options remained. In 1912 Ulster's unionists signed the Solemn League and Covenant pledging to resist Home Rule in the province by all means necessary.[9] Home Rule had been threatened for nearly thirty years, but this time was different. Prime Minister Herbert Asquith's bill had passed through the Commons, and the Lords could only delay its becoming statute for a period of three parliamentary sessions.[10] Faced with this reality, over five hundred thousand Protestants signed the Covenant, some in their own blood. They also began organizing militarily. The following year the Ulster Volunteer Force (UVF) was formed and began importing weapons illegally into the north.[11] The position of the unionists was greatly strengthened when British Army officers stationed at the Curragh Camp let it be known that, were it required of them, they would not obey orders to subdue the north in the event of an insurrection.[12]

Meanwhile, the UVF example led to parallel militarization in defense of Home Rule beginning in earnest in the south. Two paramilitary organizations came into being in 1913, James Connolly's Irish Citizen Army (ICA) and, later on in the year, the Irish Volunteers. Both organizations demonstrated the increasing militancy in the two most powerful currents in the politics of the south of Ireland: socialism and nationalism.[13] By the summer of 1914, with guns stashed all over the island and hundreds of thousands of men drilling for and against Home Rule, it was clear that the seemingly modest nationalist ambition of a degree of political self-determination within the British Empire had led Ireland to the verge of civil war.[14] The outbreak of European conflict in August 1914 served to thwart the ambitions of the more moderate nationalists, while it left those with more radical agendas convinced that constitutional politics had been tried and found wanting.[15]

Home Rule was on the statute book by the time World War I broke out, although its implementation was postponed for the duration of the conflict.[16] The crisis over a Dublin parliament had tested the allegiance of both nationalists and unionists, and World War I provided both with the opportunity to show their true loyalty to the Crown. The leaders of Ulster Unionism, Sir Edward Carson and Sir James Craig, at once promised thirty-five thousand men for the war effort. The UVF was absorbed en masse into the army and kept together as one unit, forming the Thirty-Sixth Ulster Division.[17] The Nationalist leader, John Redmond, in the flush of victory after finally seeing Home Rule jump the last legislative hurdle, also urged Irishmen to enlist in the British forces and support the war effort against Germany.[18] Crucially, this split the Irish Volunteers into the majority who supported the recruitment, numbering approximately 180,000, and a much smaller splinter faction of probably no more than 12,000 who opposed it.[19] Redmond's mistake, like so many of his generation, was to underestimate how long the war would last. As the war progressed and Home Rule did not materialize, Redmond's authority waned, and recruitment began to peter out.[20] His other mistake was to lose the support of the Volunteers, as physical force crept from the shadows onto center stage. As Joe Lee has memorably stated, "His hand was an exceptionally weak one, simply because it had no gun in it."[21]

In 1916 it became apparent just how much notions of "loyalty" had diverged in Ireland. On Easter Monday, April 24, a group of rebels in Dublin staged a coup against British rule, while in July of the same year 5,500 men of the Ulster Division were killed on the first two days of the Somme offensive.[22] To Ulster's Protestants, the treachery of the Easter Rising stood in stark contrast to the loyalty they had shown on the battlefields of Flanders. The chances of a rapprochement between nationalists and unionists were slim before 1916. After it they were nonexistent.[23]

The Easter Rising took most people by surprise, not least much of the British garrison in the city. Nevertheless, the republicans who had seized control of the General Post Office (GPO) and other key buildings were defeated and rounded up within a matter of days. Initial public reaction to the Rising among the Irish in general was hostile, particularly among those Dublin housewives queuing up outside the GPO to collect British Army separation allowances only to be told that under the new "Irish Republic" such payments were now void.[24] The Rising was a strategic failure, but the ill-judged British response turned it into a victory. The execution of fifteen of the ringleaders, including Connolly and Patrick Pearse, over a period of ten days marked the start of a change in public opinion toward the Rising.[25] During the years that followed, the more militant Sinn Féin rapidly and completely supplanted the Home Rule Party as nationalist aspirations moved from limited self-determination to full independence.[26] This change is very clear from the election results for 1918, shown in figure 6.5, which show that Sinn Féin won almost every seat in the twenty-six-county area. The exceptions were Waterford City and East Donegal, which went to the Nationalists, and Rathmines in Dublin's Anglo-Irish south side and the University of Dublin, which went to the Unionists. Conversely, north of what was to become the border, Sinn Féin only won three seats: Londonderry City, Northwest Tyrone, and South Fermanagh. The Nationalists

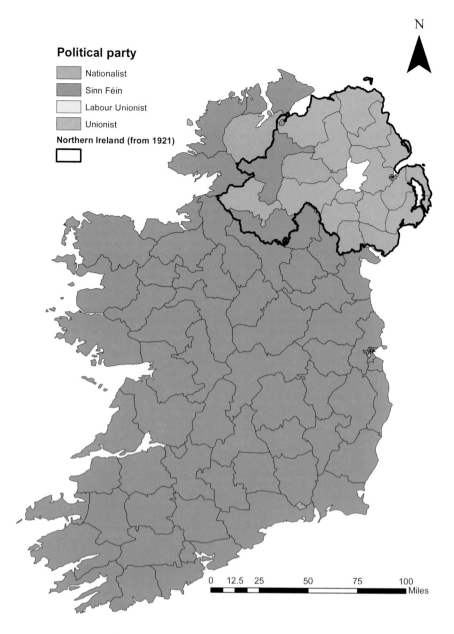

Fig. 6.5. Results of the 1918 Westminster elections. Cork City was a two-seat constituency in 1918. Sinn Féin won both seats.

Political party

- Nationalist
- Sinn Féin
- Labour Unionist
- Unionist

Northern Ireland (from 1921)

N

Party	Seats
Labour Unionist	3
Nationalist	6
Sinn Féin	73
Unionist	22

won four seats in the north, compared to only two in the south, while the remaining seats went to the Unionists and Labour Unionists.

In 1919 hostilities between the Irish Republican Army (IRA) and British forces escalated, and during the next year 192 policemen and 150 soldiers were killed by republicans.[27] Meanwhile, the British responded not only militarily but legislatively by the introduction in 1920 of the Government of Ireland Act, which finally brought Home Rule into force with separate parliaments in both Belfast and Dublin. The results of the subsequent elections to both parliaments in 1921 are shown in figure 6.6, which shows again

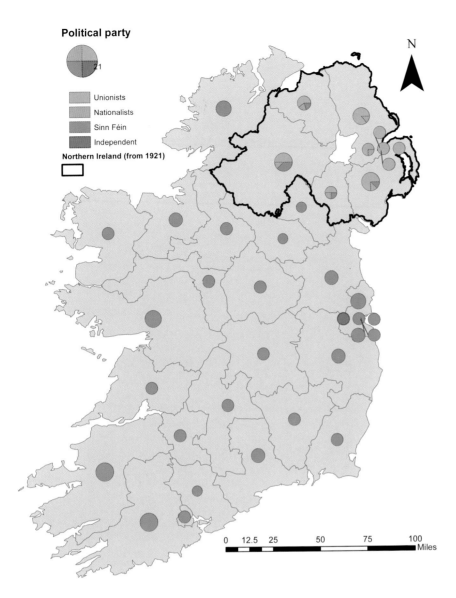

Political party

21

- Unionists
- Nationalists
- Sinn Féin
- Independent

Northern Ireland (from 1921)

N

Fig. 6.6. Results of the 1921 Northern Ireland and Southern Ireland elections. Circle size is proportional to the number of seats.

0	12.5	25	50	75	100

Miles

Party	Seats
Independent	4
Nationalist	6
Sinn Féin	131
Unionist	40

that in the southern twenty-six counties, Sinn Féin won almost every seat. The six counties that were to become Northern Ireland were dominated by Unionists, who won forty of the fifty-two seats. Sinn Féin and other nationalists won six seats each in the north, suggesting that the popularity of these two groups in relation to each other had not changed much since 1918.

The Government of Ireland Act was ignored by Sinn Féin, which by now had its own shadow administrative and legal structures in place and was operating as the de facto government of Ireland.[28] The conflict progressed with increasing bitterness as regular British troops were reinforced by the ill-disciplined "Black and Tans" until a truce was finally called in July 1921.[29] Protracted negotiations led to the signing of the Anglo-Irish Treaty in December.

Fig. 6.7. Results of the 1922 Northern Ireland and Saorstát Éireann elections. Circle size is proportional to the number of seats.

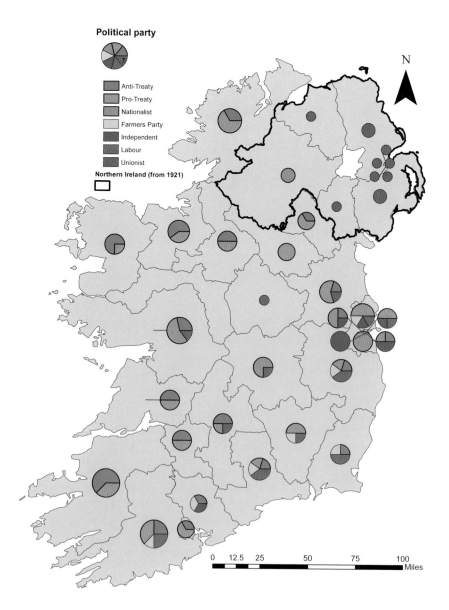

Political party

- Anti-Treaty
- Pro-Treaty
- Nationalist
- Farmers Party
- Independent
- Labour
- Unionist

Northern Ireland (from 1921)

Party	Seats
Farmers' Party	7
Independent	10
Labour	16
Antitreaty	34
Protreaty	53
Nationalist	2
Unionist	10

The treaty split nationalist Ireland not primarily because it conceded the exclusion of the six counties of Northern Ireland but because it required members of the new Irish parliament (Dáil) to swear an oath of allegiance to the British monarch and it failed to deliver a full Irish Republic.[30] The treaty was ratified by the Dáil and had the clear support of the majority of people in the twenty-six counties. The 1922 elections were effectively a referendum on the treaty.[31] Again, as figure 6.7 shows, clear geographies can be found. The antitreaty vote was concentrated in the west and south-

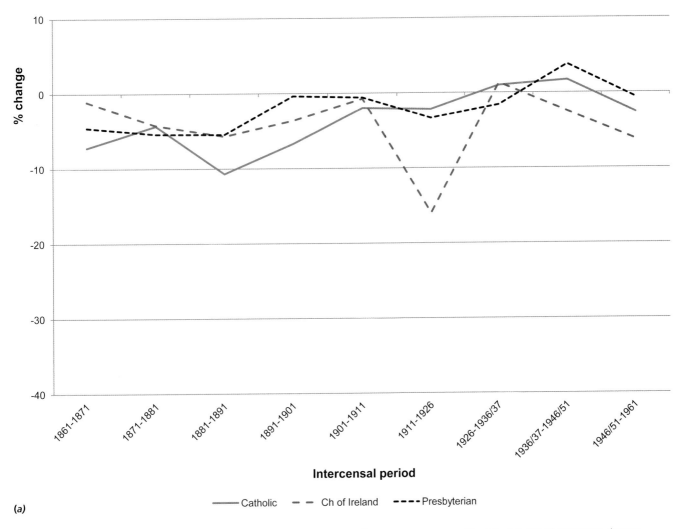

Fig. 6.8a. Intercensal percentage changes in the three major religions between 1861 and 1961 in (a) the Northern Ireland area. (*Continued on the next page*)

west, while the east, particularly around Dublin, was strongly protreaty. The opposition in the southwest meant that this was the area that saw the most intense fighting of the Civil War and some of its worst atrocities, such as the Ballyseedy Massacre, in which nine antitreaty supporters were tied to a landmine, which was then detonated—an example of the sort of summary justice not uncommon during the period.[32] Despite the fact that the protreaty faction won the 1922 election, Eamon De Valera and those opposed to the treaty argued that as the Republic had not yet been realized, the popular will of the people could not yet be respected.[33] Over the next year many of those who had come to symbolize the new state, most notably Michael Collins, were killed while fighting for the future of independent Ireland, fighting that was remarkable in its ferocity and bitterness.

The fact that such deep division could exist among nationalists concerning the extent and nature of the independence they had achieved showed how unrecognizable the island had become in the space of just eleven years. Any sense of achievement had to be balanced by a pervasive disappointment. Nationalists in the south considered their independence to be a form of dominion status within the British Empire that went much further than they wanted in 1914 but fell short of their expectations in 1921 or 1922.[34] Ulster's Protestants finally had a form of autonomy forced upon them that they had not wished for in the first place, and they had sacrificed their coreligionists in the south to "Rome rule" in the process.[35]

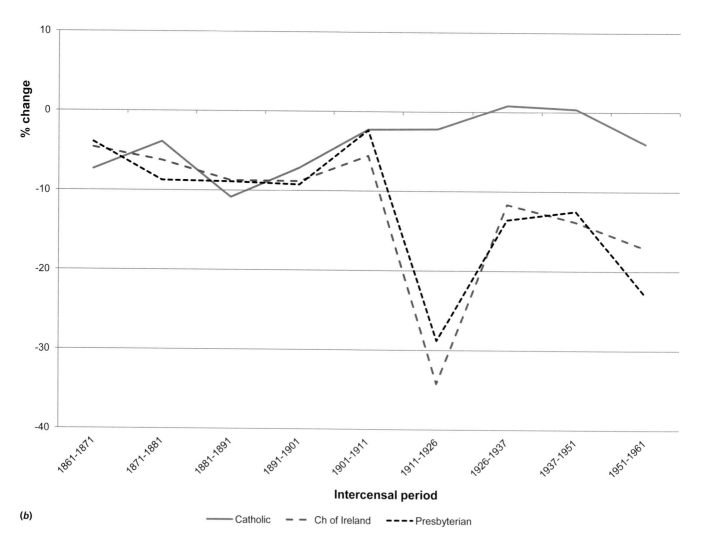

Fig. 6.8b. Intercensal percentage changes in the three major religions between 1861 and 1961 in (b) the Republic of Ireland area.

Religious Divergence: Moving Apart or Moving Away

From the previous section it is clear that religious identity markedly affected politics in this period and that this played out geographically, leading to the island being partitioned on largely sectarian grounds. Clearly, however, Partition would also have the potential to influence the geographies of religion, as many people would be on the "wrong" side of the newly created border and the increasingly entrenched and sometimes violent political attitudes that went with it.

Figure 6.8 shows that up to 1911 Ireland's three main religions followed broadly similar demographic trends both north and south of what was to become the border. Figure 6.8a shows the intercensal population change in the religions in the Northern Ireland area across the century from 1861. After the Famine all three groups had experienced population decline, which was more marked among Catholics, whose population took longer to stabilize. Figure 6.8b shows the same information for the Republic of Ireland area over the same period. Prior to 1911 it shows a pattern that is similar to the north but more pronounced in that all three groups show a decline, which appears to have stabilized by the 1900s. However, between 1911 and 1926 the situation changed dramatically as the two major Protestant denominations experienced massive levels of population loss: the Church of

Ireland population in the south collapsed by nearly 35 percent, while the number of Presbyterians fell by 27 percent.

Maps of the Catholic population are presented in figure 6.9, and the individual Protestant denominations are shown in figure 6.10a and 6.10b. Percentage change over this period for the three denominations is shown in figure 6.11. Table 6.1 shows the changing populations of the religions north and south of the border. Figure 6.11b, in particular, shows that there was a major loss of the Church of Ireland population from south of the border. Even if the border had not been included on this map, its location would be apparent, as south of it most districts with a significant Church of Ireland population showed a decline of more than 20 percent, while north of it few districts showed declines at this level. For Presbyterians this pattern is less marked, as there were few districts south of the border that were more than 5 percent Presbyterian. Figure 6.11a, however, suggests that there is little evidence that Partition had a major impact on the distribution of the Catholic population. The trends on this map are consistent with earlier maps, with patterns of loss in the west and gain in the east, particularly in the urban centers, which, despite Partition, still included Belfast. This pattern of ongoing change caused by economic developments is also perhaps apparent in the two maps of the Protestant population, which also show gain in the urban centers, particularly Belfast but also Dublin in the case of the Church of Ireland, and loss in the more rural areas, although this is less pronounced north of the border than it is in the south.

The changes over this period led to Ireland's religions becoming more geographically isolated from each other. The index of dissimilarity rose from 55.7 percent in 1911 to 65.1 percent in 1926. The trend over the previous three decades had been for this index to rise at approximately 1 percent per decade, but over the 1911–26 period this rose to the equivalent of almost 5 percent per decade. Isolation indexes for both Catholics and Protestants also rose significantly faster than would be expected given the overall population changes. For Protestants the isolation index rose from 46.4 percent to 48.3 percent, a rise that is 2.4 percentage points higher than would be expected, while Catholics increased from 83.4 percent to 85.7 percent, a rise that was 0.8 percentage point above what would have been expected. This shows that even working with county-level data, Partition caused an unprecedented rise in the extent to which Catholics and Protestants lived farther apart from each other.

Table 6.1 summarizes the impact of these changes. The major loss of Protestants from the south is clear. What is particularly interesting, however, is the lack of significant change in the north. This tells us two things. First, while Protestants clearly left the Free State in large numbers, Catholics did not leave Northern Ireland to any discernible extent. Second, while the non-Catholic population of the south fell by 115,000, the non-Catholic population of the north only rose by 23,000, suggesting that rather than there being a major migration from south to north, large numbers of Protestants left Ireland altogether. Some of this may be explained by British military and government personnel leaving, but the fact that most of this loss seems to have occurred in rural areas while the Church of Ireland population of Dublin actually increased suggests that the British loss was not a major factor.

Fig. 6.9. Catholic populations in (*a*) 1911 and (*b*) 1926.

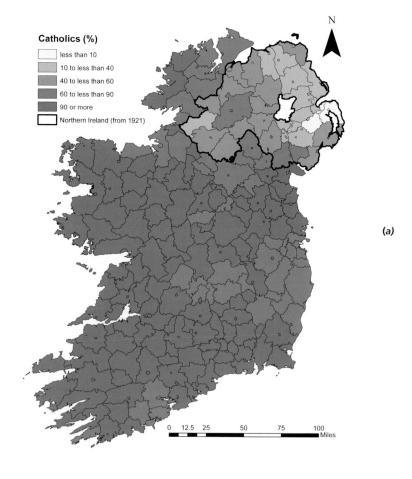

Catholics (%)

less than 10
10 to less than 40
40 to less than 60
60 to less than 90
90 or more
Northern Ireland (from 1921)

0   12.5   25   50   75   100
Miles

(*a*)

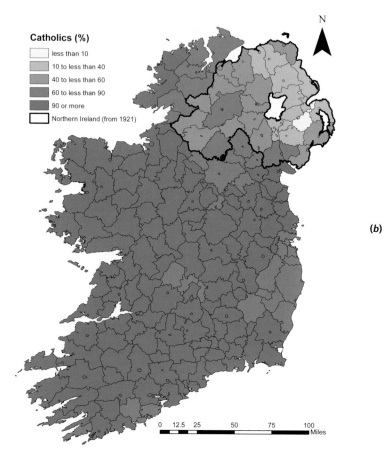

Catholics (%)

less than 10
10 to less than 40
40 to less than 60
60 to less than 90
90 or more
Northern Ireland (from 1921)

0   12.5   25   50   75   100
Miles

(*b*)

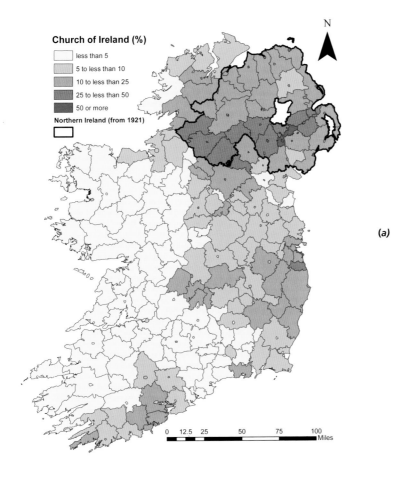

Church of Ireland (%)

less than 5
5 to less than 10
10 to less than 25
25 to less than 50
50 or more

Northern Ireland (from 1921)

N

(a)

0   12.5   25   50   75   100
Miles

Fig. 6.10a. Church of Ireland populations in 1911 and 1926. (*Continued on the next page*)

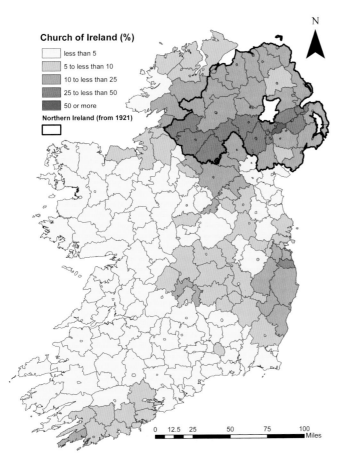

Church of Ireland (%)

less than 5
5 to less than 10
10 to less than 25
25 to less than 50
50 or more

Northern Ireland (from 1921)

N

0   12.5   25   50   75   100
Miles

Fig. 6.10b. Presbyterian populations in 1911 and 1926.

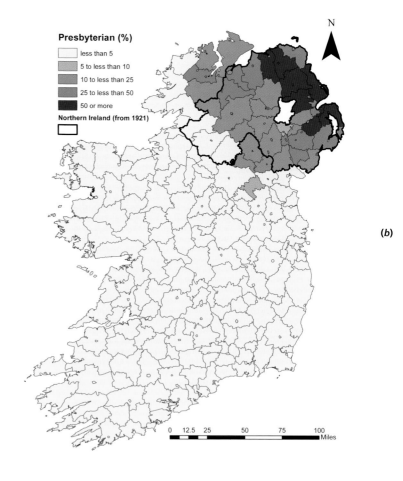

Presbyterian (%)

- less than 5
- 5 to less than 10
- 10 to less than 25
- 25 to less than 50
- 50 or more

Northern Ireland (from 1921)

N

0 12.5 25 50 75 100
Miles

(b)

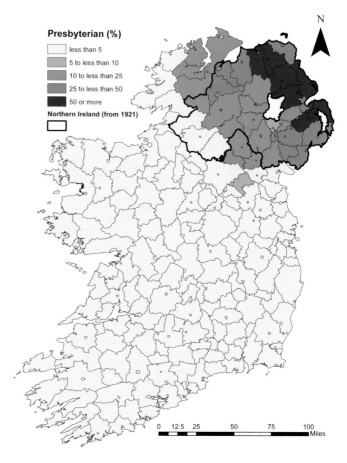

Presbyterian (%)

- less than 5
- 5 to less than 10
- 10 to less than 25
- 25 to less than 50
- 50 or more

Northern Ireland (from 1921)

N

0 12.5 25 50 75 100
Miles

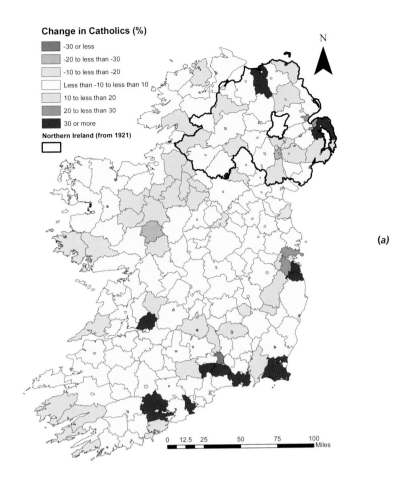

Change in Catholics (%)

- -30 or less
- -20 to less than -30
- -10 to less than -20
- Less than -10 to less than 10
- 10 to less than 20
- 20 to less than 30
- 30 or more

Northern Ireland (from 1921)

N

0 12.5 25   50   75   100
Miles

(a)

Fig. 6.11a and 6.11b. Change in the populations of (a) Catholics, (b) Church of Ireland, and (c) Presbyterians (see next page), Districts with less than 5 percent of their population being from a denomination in 1911 have not been shaded.

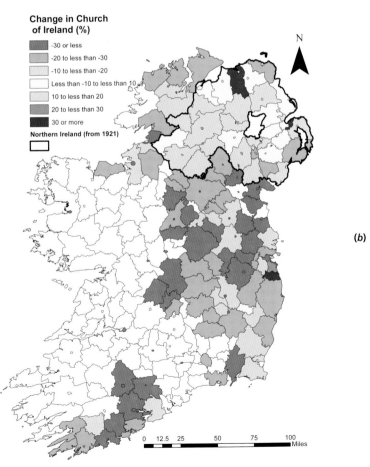

Change in Church of Ireland (%)

- -30 or less
- -20 to less than -30
- -10 to less than -20
- Less than -10 to less than 10
- 10 to less than 20
- 20 to less than 30
- 30 or more

Northern Ireland (from 1921)

N

0 12.5 25   50   75   100
Miles

(b)

Fig. 6.11c. Change in the populations of (c) Presbyterians, 1911–26. Districts with less than 5 percent of their population being from a denomination in 1911 have not been shaded.

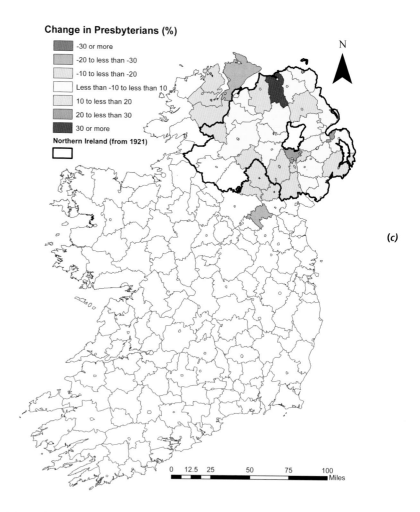

Change in Presbyterians (%)

- -30 or more
- -20 to less than -30
- -10 to less than -20
- Less than -10 to less than 10
- 10 to less than 20
- 20 to less than 30
- 30 or more

☐ **Northern Ireland (from 1921)**

N

(c)

0 12.5 25 50 75 100 Miles

Table 6.1. Population changes in the six northern counties and twenty-six southern counties, 1911–1926

		Pop.		Change		% of Pop.	
		1911	1926	*n*	%	1911	1926
North	Cath.	429,089	420,428	−8,661	−2.0	34.5	33.5
	Non–Cath.	813,120	836,133	23,013	2.8	65.5	66.5
	Col	325,207	320,001	−5,206	−1.6	26.2	25.5
	Pres.	389,913	393,374	3,461	0.9	31.4	31.3
	Pop.	1,242,209	1,256,561	14,352	1.2	100.0	100.0
South	Cath.	2,811,810	2,751,269	−60,541	−2.2	89.3	92.6
	Non–Cath.	335,705	220,723	−114,982	−34.3	10.7	7.4
	Col	249,281	164,167	−85,114	−34.1	7.9	5.5
	Pres.	45,566	32,478	−13,088	−28.7	1.4	1.1
	Pop.	3,147,515	2,971,992	−175,523	−5.6	100.0	100.0

Cath.: Catholics; Non-Cath.: non-Catholics; Col: Church of Ireland; Pres.: Presbyterians; Pop.: total population.

Figure 6.12 shows a scatter plot that compares the non-Catholic populations of rural and urban districts in 1911 and 1926. Most districts lie on a line that runs at 45 degrees across the diagram, indicating that their non-Catholic populations in 1911 and 1926 were similar. Districts lying below this line saw a decrease in their non-Catholic populations; those lying above it saw their non-Catholic populations increase. Very few districts

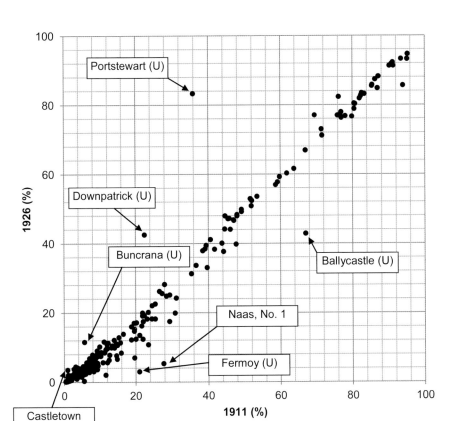

Fig. 6.12. Scatter plot of non-Catholic rural and urban district populations, 1911 and 1926.

south of the border saw growth in their non-Catholic population. Two that did were Castletown in the extreme southwest and Buncrana Urban in County Donegal. The likely reason for this growth is that these districts contained "treaty ports," naval bases that remained under British control as part of the peace settlement.[36] The rural district of Naas, No. 1, on the other hand, shows a substantial decline in its non-Catholic population between 1911 and 1926, and this decline can be attributed in large part to the area's strategic importance as the British Army withdrew from its largest barracks in Ireland on the Curragh.[37] The same factor also underlies the change in Fermoy, another important military town in north County Cork. The scatter plot also contains a number of outliers in Northern Ireland, such as Ballycastle (U), Portstewart (U), and Downpatrick (U).[38] These unusual figures can be explained by changes to boundaries between the two dates rather than to substantial shifts in the religious population.

The direct causes of the dramatic loss of Protestant population from the Free State are difficult to discern. Violence and intimidation may well have played a part, but violence took place on both sides of the border and affected both communities, not all of which seemed to show any population loss. During 1920 and 1921 the IRA conducted operations along the border to destabilize the new Northern Ireland state that was about to emerge. In Belfast the sectarian problem was compounded by economic depression in the postwar downturn. The Belfast "Pogroms" of the early 1920s saw thousands of Catholics driven from their jobs and homes, some to become refugees across the border. Many Protestants were also displaced during this period as the city's religious boundaries hardened, undermining the use of the term "pogrom," although Catholics certainly suffered disproportionately. Worst of all, during the first six months of 1922 intercommunal rioting led to the deaths of 171 Catholics and 93 Protestants, often in the most horrific of circumstances.[39] In all cases, however, the Belfast area still shows some of the largest growth rates of anywhere in Ireland, suggesting that while these events were clearly highly traumatic, at the scale we are working at their demographic impact was limited.

South of the border, violence against Catholics in Belfast is likely to have been an aggravating factor in the Dunmanway Massacre of ten Protestants in west Cork during April 1922. According to contemporary reports, the massacre prompted a rapid exodus of Protestants from the county and presumably out of their country, many never to return.[40] Figure 6.11b confirms that west Cork did indeed show widespread losses of its Church of Ireland population. Lower-level violence and intimidation were also occurring. "Land grabbing," as it became known, was a major issue during this period, particularly in border areas, but it affected rural Protestants throughout the Free State area.[41] While the victim profile was clearly sectarian, the motivation was more often local jealousy and personal animosity rather than overt religious hatred. Nevertheless, the result was the same, and it added to the sense of fear and isolation that many southern Protestants felt, at least in the early stages of independence.[42]

Certainly, political events profoundly altered Ireland's religious geographies during this period, but there are two aspects of the crisis that are worthy of particular note. First, there is the domino effect of events, whereby an actual or perceived sectarian attack in one part of the island could have profound consequences in the form of retaliation on an embattled religious minority hundreds of miles away. There appears to be stronger temporal links between instances of violence than spatial ones. Such appears to have been the case with the Belfast Pogroms and Dunmanway Massacre of the spring of 1922, and although there is debate over how overtly sectarian the motives behind the west Cork attack were, there is little doubt that its timing was directly influenced by events north of the border, coming only a couple of days after some of the most serious attacks on Catholic areas of Belfast. In turn, the IRA's border attacks are believed to have led to the sectarian assault on Catholic areas of Belfast.[43] This assault was dubbed by the Catholic bishop of Down and Connor, Joseph MacRory, as the "doctrine of vicarious punishment," wherein "the Catholics of Belfast are made to suffer for the sins of their brethren elsewhere."[44]

Second, the everyday realities of religious violence and expulsion were not an accurate reflection of higher-level political thinking. There was no policy to expunge the Protestant population of the Irish Free State; quite the opposite, as the new government tried to prove its inclusivity through weighted representation in the legislature and laws designed to protect Protestant freedoms.[45] North of the border, the Catholic minority, hostile as they were to the creation of the Northern Ireland state, were viewed with intense suspicion.[46] However, the six-county territorial settlement was recognition of the fact that there would always be a substantial Catholic population within Northern Ireland's boundaries. From Northern Ireland's very inception, the ruling unionist government showed a powerful understanding of the meaning of boundaries and of its willingness to manipulate them in order to solidify Protestant control of the state. Senior unionists were also intensely concerned about the danger posed by the IRA as well as by loyalist gunmen who were similarly out of control. It was a desire to rein in these elements that led to the formation of the Ulster Special Constabulary (USC) out of the paramilitary UVF.[47]

Formalizing the Division: Partition

The gestation of Partition as a serious solution to the conflict in Ireland was remarkably short.[48] The last section of this chapter will briefly examine how the 1921 arrangement was arrived at as well as some of the alternative scenarios based upon the 1911 census data.

Figure 6.13 shows the recognizable six-county Northern Ireland that was adopted by the British government. As table 6.2 shows, this solution was perhaps the best that the unionist administration to be created in Belfast could have hoped for, as it was the largest area that could be created on a county framework that had a sustainable Protestant majority, having a Catholic population of only 34 percent based on the 1911 census figures.[49] Unionists were perhaps fortunate to achieve a six-county solution, since, based on the aggregate unit of the county, there was no religious demographic justification for the inclusion of the counties of Fermanagh and Tyrone, neither of which had ever had a Protestant majority population.

The fact that Fermanagh, Tyrone, and Derry City were not to become part of the Free State is indicative of a number of important points. First, it reflects the aforementioned view that Partition was of less concern to most nationalists than the oath or Ireland's status vis-à-vis the Crown and the concession of the British monarch as titular head of state. Second, and with specific regard to Partition, nationalists set great store by the Boundary Commission, which was to be established by the British government and which would supposedly resolve once and for all the territorial issues. Not only the Irish but also the British believed that the formation of this commission would logically lead to the restitution of at least the counties of Fermanagh and Tyrone, with their clear nationalist majorities, to the Irish Free State.[50] Underpinning this was a belief that after the border was realigned the northern state would no longer be economically viable and would seek reabsorption into the Irish Free State.[51] Such a naive viewpoint reflected the ignorance of economic realities in Ireland and the fact that northeast Ulster had an extremely buoyant and outward-facing industrial base far more reliant on the cities of Britain than the meadows of Fermanagh. In the end the logic of this reading would not be tested, as the Boundary Commission did not come up with the expected results. Instead of returning large parts of Northern Ireland to the Free State, the commission suggested piecemeal changes, with fewer than twenty-five thousand people and two hundred square miles of territory being transferred to the south.[52]

Figure 6.13 also shows an alternative settlement based on the county framework that was advanced by nationalists in 1914. This would have seen just the four counties with Protestant majorities forming a Northern Ireland that would have had a Catholic minority of 30 percent of the population. This was never likely to be accepted by unionists. One of the arguments advanced by way of objection was that the four-county unit would have

Fig. 6.13. The Partition of Ireland creating Northern Ireland and the Irish Free State. Shaded counties are the four with Protestant majorities. Donegal, Cavan, and Monaghan are the three Ulster counties that were not included into Northern Ireland.

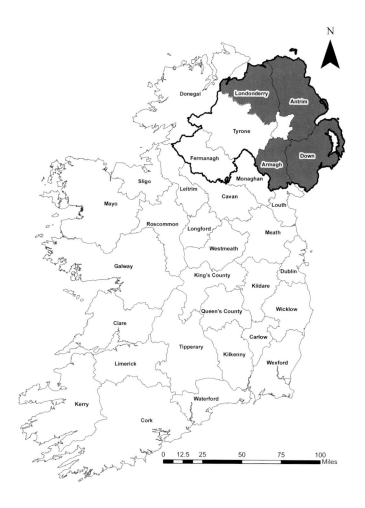

Table 6.2. The proportion of the population of Northern Ireland that would be Catholic or Protestant depending on different definitions of the new state

	Catholic	Protestant
Six-county solution	33.7	66.3
Four-county solution	30.0	70.0
Provincial settlement	42.9	57.1
Contiguous districts	26.0	74.0

been administratively unsustainable rather than economically so. It would not have justified the expenditure on government structures that the geographical entity would have required.[53] This is an interesting argument, as demographically there were British dominions such as Newfoundland and New Zealand with smaller populations than the four-county area. David Lloyd George, the British prime minister, favored the provincial settlement, which would have seen all nine counties of the province of Ulster forming Northern Ireland.[54] Given the frequency with which Ulster and its mythological symbols such as the heraldic Red Hand were appropriated by unionism, one might have been forgiven for thinking that residents of Ulster would naturally have been in favor of this logical territorial settlement. Initially this was indeed the case, but by 1914 Carson had swung against the provincial solution, and table 6.2 helps explain why. Five of Ulster's nine counties had Catholic majorities, which in total would have contributed

43 percent of Northern Ireland's population. The fear among unionists was that high Catholic birthrates would have led to a loss of the Protestant majority within a generation. However, the problem with the six-county solution was that it resulted in the remaining three Ulster counties, all of which had significant Protestant minorities, becoming part of the Free State. Lee condemns this, calling the desertion of those southern border Protestants "one of the basest betrayals of the period."[55] With perhaps more empathy for the position of Carson, Michael Laffan has argued that the leaders of unionism were simply trying to secure the largest area they could for their people, in line with contemporary European nationalist movements after World War I.[56]

A final possible solution would have been to have moved away from the use of counties as the basic unit on which Partition would be based and define Northern Ireland using contiguous districts with a simple non-Catholic majority. This would have left an emasculated state that would have included all of Antrim but only the northern parts of Armagh, Down, and Londonderry, although the resulting state would, as table 6.2 shows, have been one with a Protestant majority of three to one. The main centers of Catholic population under this counterfactual dispensation would have been in Belfast and in Derry City, where in the latter Catholics were actually a majority of the city's population. Henry Patterson has identified border unionists as being a particularly reactionary and militant element within Ulster unionism during the post–World War II period.[57] This can be understood to have arisen at least partially from the demographic and geographical vulnerability many unionists in border areas of Northern Ireland felt. A more demographically secure Northern Ireland than the six-county arrangement might have acted to alleviate the "siege mentality" many unionists still lived with.[58] On the other hand, increasing the Protestant majority by excluding areas where large numbers of Catholics lived would have resulted in a much smaller Northern Ireland geographically. Hypothetical analyses like these only serve to underline the power of spatial scale in determining potential outcomes.[59] It would also have left more of Ulster's Protestants living in the Free State.

Conclusions

Through the nineteenth century, economic and social divisions increased, and these divisions frequently fell along religious lines. There were clear, closely related geographies to all three of these divides. Home Rule added a fourth tier to this: politics, more precisely, the politics of how Ireland was to be governed. Against the backdrop of the First World War and the Easter Rising, this politicization became increasingly extreme. Nationalist moves toward Home Rule within the British Empire were replaced with demands for full independence that would not even accept the British monarchy as titular head of state. Unionism was implacably opposed to Home Rule, and both sides became increasingly prepared to use violence to achieve their aims.

The spatioreligious effects of Partition are interesting. The Free State, which, at government level at least, was not opposed to Protestantism, nevertheless saw a large-scale loss of Protestants, many of whom must have left the island altogether. In Northern Ireland, where there was deep suspicion of Catholics at government and more popular levels, there was no such loss; indeed, despite Partition and the violence in Belfast, the Catholic population of east Ulster continued to grow unabated. One reason for this is likely to have been that economic prerogatives overrode politics and even intimidation and violence. While Belfast continued to offer jobs, both Catholics and Protestants would move there. Conversely, the lack of economic opportunity in the Free State, combined with doubts about its political future, not to mention ongoing lawlessness and violence in some areas, may well have contributed to large numbers of Protestants leaving.

The consequence was that the British-facing, economically successful part of Ireland remained part of the United Kingdom. This area had a Protestant majority; however, the Catholic minority made up about a third of the population. The Free State was determined to move away from Britain politically and economically. Its Protestant minority was far smaller than the Catholic minority in the north and was reduced in both numbers and geographical spread by Partition. By 1926 only Dublin, Wicklow, and the three Ulster counties that lay outside Northern Ireland were more than 10 percent Protestant. Religiously, therefore, the Free State was much more homogeneous than Northern Ireland, and the Protestants there were concentrated in a smaller number of areas than in the north.

The challenge for both parts of the island was how to be economically successful and socially harmonious. For the north, economically, this meant a continuation of the policies that had served it well. Religiously, the fear of becoming part of the Free State and the relatively large Catholic minority would present a challenge to the north's ability to become socially cohesive, especially as the Catholic minority felt, and frequently was, marginalized and discriminated against. The Free State would have to reinvent itself economically, as blaming British misrule was no longer an option.

Division and Continuity, 1920s to 1960s 7

The Boundary Commission of 1925 confirmed the territorial settlement of Partition. Ireland would remain divided. In many ways Northern Ireland and the Irish Free State had the same central problem at the start of this period: the 1921 treaty had created two states, but it had not created two nations.[1] Religious geographies had determined the spatial extents of both jurisdictions, but the choice for both the north and south was how to forge their own identities and the extent to which these identities would be defined by the sectarianism of their geneses. The new formalized division of Ireland was, as we have seen, about more than just religion—it closely reflected the social and economic divisions of the island as well. A second challenge was thus to develop their separate economies. A final question was whether Partition would mark a new beginning for Ireland or whether it would simply continue the trends that had been developing since the mid-nineteenth century.

The Politics of Population

As we will see, the political division of Ireland did not have a profound effect on its diverging demographic trends in the short to medium term. The decline in the Irish Free State continued apace, while the increase in Northern Ireland's population persisted unremarkably but consistently. What Partition did change, however, were attitudes toward population decline. In the past it had been possible to blame population failure on the vampire nation across the Irish Sea, a nation that was draining the lifeblood from the Irish nation through a mix of willful economic mismanagement and the denial of self-determination, which would give the Irish the power to remedy the problem.[2] The establishment of the Irish Free State removed these excuses and gave the Irish both the power and the responsibility to tackle the single greatest social issue facing them.

Around the turn of the century there was a powerful current in unionist thinking, heavily influenced by the work of Max Weber, that saw the economic successes of northeast Ulster as being directly influenced by the dominant Protestant character of the region.[3] There was nothing unusual about the employment of racial modes of explanation for economic inequalities in this period, not least because nationalists were using the same arguments to explain the innate "passion" and "hospitality" of the Gaelic, Catholic Irish.[4] So there was nothing new about competition between the two parts of the island. The difference was that in the past, northeast Ulster, although distinctive, had still been part of a larger Irish geographical and political entity. After Partition this was no longer the case, and the issue of demography became an important and symbolic political football to be kicked back and forth across the border. Given the failure of independent Ireland to stem its population decline throughout almost all of this period, the issue would at times be used as an important propaganda tool for Northern Ireland.

Figure 7.1 shows the distribution of people born in one state but living in the other, in other words, people born in the south living in Northern Ireland and people born in Northern Ireland and living in the south. There were consistently higher levels of people living in Northern Ireland who had been born in the south than the opposite, although the extent to which this is because of the larger population base of the south is hard to discern. What is perhaps more important, however, is the trend over time, and this shows a decline in the percentage of people born in the south and living in the north after Partition. This runs counter to popular thinking in unionist circles and was also the subject of academic debate in the south.[5]

Figure 7.2 shows the change in population density over this period. It is clear that in 1926 most of the island was already extremely sparsely populated, as the long-term population decline following the Famine continued throughout the south. A large majority of districts had fewer than 0.5 people per acre. However, even from that exceptionally low base, a further decline in the rural population is evident from the time series shown in figure 7.2, as more and more districts fell below 0.1 people per acre. The decline starts on the Atlantic seaboard in the coastal districts of Mayo, Galway, and Kerry. A large inland area on the west of the Central Plain is also evident. These areas of decline spread during the course of the next thirty-five years as the population of the west and center of the island continued to drop. By 1961 a substantial part of the island's population density had fallen below the 0.1 people per acre mark. It is also clearly apparent that this phenomenon of rural decline had a far more serious impact in the Republic of Ireland area than in Northern Ireland. During the period 1926 to 1961, Northern Ireland largely maintained the rural population base it started with, although a clear exception exists in the district of Fermanagh, to which the disease of rural depopulation had clearly spread.

The rural decline that figure 7.2 identifies does not appear to be offset by any major geographical spread in the urban centers, although some limited suburban growth is evident with the increase in densities in areas such as Lisburn, west of Belfast, and Lucan, west of Dublin.

Figure 7.3 builds on this by summarizing patterns of growth between 1926 and 1961. It shows that the heaviest declines occurred in the northwest of the Republic of Ireland area but also stretched right along the border to the Irish Sea in the east. Northern Ireland also suffered some regional population decline in the border region, although it was not of the same magnitude as that which occurred in the south; indeed, the border appears to mark a clear line between lower rates of loss in the north and higher rates in the south. A much larger proportion of Northern Ireland districts saw either population stability or increase. Much of County Clare and the peninsulas of the southwest also saw severe decline during these thirty-five years.

Figure 7.3 also gives a much clearer pattern of Ireland's accelerating process of urbanization. Areas of stability and growth are all in the hinter-

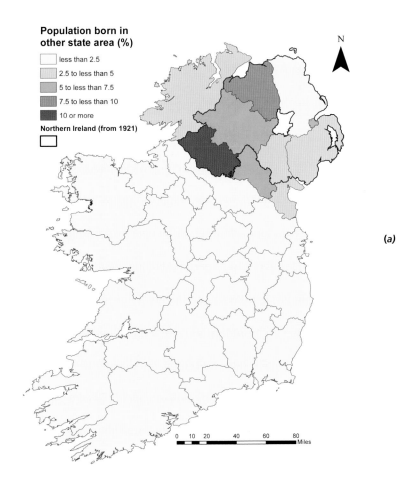

Population born in other state area (%)

- less than 2.5
- 2.5 to less than 5
- 5 to less than 7.5
- 7.5 to less than 10
- 10 or more

Northern Ireland (from 1921)

N

0 10 20 40 60 80 Miles

(a)

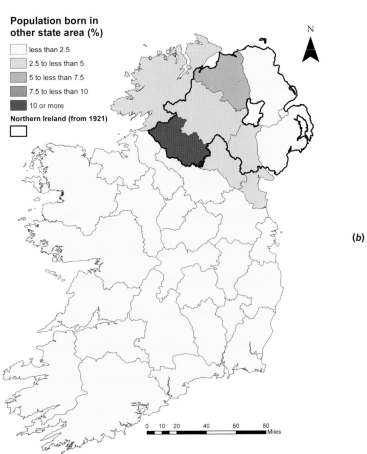

Population born in other state area (%)

- less than 2.5
- 2.5 to less than 5
- 5 to less than 7.5
- 7.5 to less than 10
- 10 or more

Northern Ireland (from 1921)

N

0 10 20 40 60 80 Miles

(b)

Fig. 7.1. Percentage of the population at the county level born in the northern six counties of Ireland and resident in the southern twenty-six counties, or vice versa, in (a) 1926 and (b) 1946/51.

Fig. 7.2. Population density at rural and urban district levels: (*a*) 1936/37, (*b*) 1946/51, and (*c*) 1961. See figure 6.2 for earlier maps in this series.

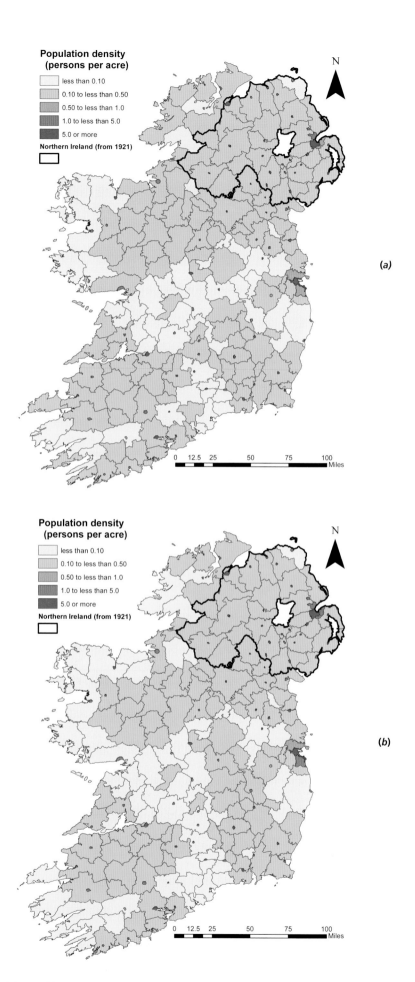

(*a*)

(*b*)

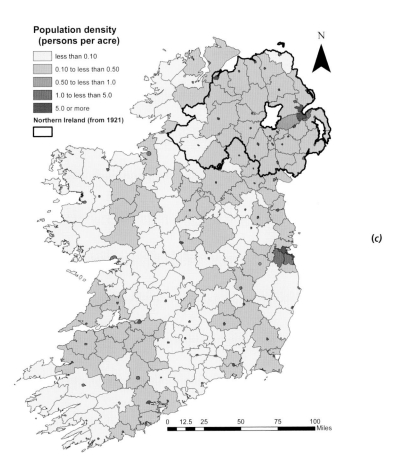

**Population density
(persons per acre)**

less than 0.10
0.10 to less than 0.50
0.50 to less than 1.0
1.0 to less than 5.0
5.0 or more

Northern Ireland (from 1921)

(c)

0   12.5   25   50   75   100
Miles

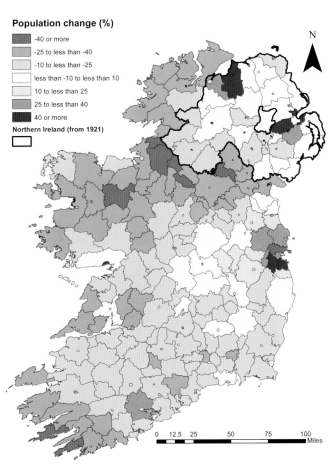

Population change (%)

-40 or more
-25 to less than -40
-10 to less than -25
less than -10 to less than 10
10 to less than 25
25 to less than 40
40 or more

Northern Ireland (from 1921)

0   12.5   25   50   75   100
Miles

Fig. 7.3. Percentage of population change at rural
and urban district levels between 1926 and 1961.

lands of major urban settlements, most notably surrounding Belfast and Dublin but also around Cork, Londonderry/Derry, Galway, and to a lesser extent Limerick and Waterford. The severe decline in the rural district of Dublin North may appear anomalous given its proximity to such a major city, but this likely can be explained by boundary redefinitions as a result of the expansion of the Dublin urban area.

It is clear that the population crisis posed a major problem for the successive administrations in the Irish Free State and the Republic and also that whatever strategies they developed to deal with the crisis were largely ineffective. What did these administrations do, and why did such measures fail to stem the population hemorrhage?

Economics, Industrialization, and Identity

Ireland's population ills had long been blamed on its failure to industrialize, and the continued success of the Belfast region in the first part of the twentieth century only served to illustrate this failure.[6] Independent Ireland's economic character from Partition to the early 1950s can be divided into three periods. The first period was characterized by economic conservatism and consolidation under the first protreaty Cumann na nGaedhael administration. This period was followed by the more aggressive fiscal policies of Eamon De Valera's early Fianna Fáil governments from 1932. The last discrete period that can be identified is that commencing with the outbreak of World War II—or the Emergency, as it was known in the Free State—which lasted through to the war's aftermath.

The protreaty Cumann na nGaedhael Party formed the first government in the Irish Free State. The country was in a desperate state economically, and the Civil War that had just been won would live on to define the political culture of independent Ireland.[7] The Cumann na nGaedhael administration was not marked out by its inventiveness. Its greatest priority in the first decade of independence was to maintain security, as republicans still posed a very serious threat to the stability of the new state.[8] The urgency of this was illustrated by the assassination of the minister for justice, Kevin O'Higgins, in 1927.[9] The administration also possessed an almost obsessive attitude toward fiscal rectitude and balancing the budget.[10] Its natural conservatism militated against making the sort of big investments a process of industrialization would require.

These ideological shackles made the establishment of the "Shannon scheme" between 1925 and 1929 all the more remarkable. This scheme built the largest hydroelectric dam the world had yet seen across the River Shannon near a village called Ardnacrusha. The Electricity Supply Board (ESB) was established out of this project as part of an ambitious national electrification program.[11] The harnessing of the Shannon was a powerful symbol of the new state's desire to forge its own modern identity, but in reality it did not mark the beginning of a new industrial departure in Ireland. If anything, postcolonial Ireland was remarkable for its economic continuities rather than its disruptions. Expert advisors recommended that the early government should do all it could to maintain the fiscal status quo: peg the currency to sterling, continue to concentrate on agriculture, continue with a market dependence on the U.K., and maintain a free-trade area so as not to rock the boat with the sort of protectionist policies that were coming into vogue across the world at the time.[12]

The problem? The status quo *was* the problem as the population drift continued. Rural life was being idealized by the state and reinforced by pervasive Catholic social teaching as the optimal form of existence, yet the people in country areas were clearly voting with their feet, as the population maps in figures 7.2 and 7.3 confirm.[13] In a sense the romantic attitude of the church and the government toward a supposed pastoral idyll was a contrary one. The previous chapters showed that the prevailing economic and demographic forces that emerged from the Irish countryside after the Famine were leading to the concentration of land and wealth in the hands of fewer people with a concomitant purging of the rural population. Yet the church and the government were merely articulating a commonly held belief that the land was a repository for the nobler aspects of the national psyche, and of course there was no light without darkness; the cities were suitably demonized, and not without good reason in many respects.[14] As late as the 1940s the life expectancy of a child born in Dublin was ten years lower than one born in the rural west of Ireland.[15]

Cumann na nGaedhael was unwilling to experiment greatly in economic terms; however, the establishment of semistate bodies like the ESB and the Irish Sugar Company was a step forward.[16] The ESB was an infrastructural and moral boost, while the founding of the Irish Sugar Company signaled a commitment to build up Ireland's agricultural base rather than to radically redesign the entire economy. If one of the early goals of the Irish Free State was to preserve an overwhelmingly agricultural society, then it is clear from figure 7.4 that its first governments were very successful in this task. Agriculture continued to be by far the most important sector across the south and the largest single source of rural employment. An ongoing commitment to agriculture was seen not only as an economic imperative but also as an ethical foundation of Irish society.[17]

The patterns from figure 7.4 contrast with those in figure 7.5, which shows the lack of manufacturing employment throughout the Republic area. Although the general patterns may be fairly clear, some interesting nuances exist within these maps. The relatively high level of manufacturing employment in County Louth may be explained in part by its significance as a center of footwear production between the wars.[18] Figure 7.3 also suggests another example of the importance of local economic circumstances. It has already been noted how the Irish midlands saw significant population decline during the period 1926 to 1961. However, a number of contiguous rural districts to the west of Dublin show stability and even modest growth during the same period. This is likely to be at least in part a result of the establishment of Bord na Móna, or Turf Board, in 1948.[19] The Turf Board was set up to harvest the peat reserves located in the large bogs of Ireland's Central Plain. According to Daly, the establishment of Bord na Móna had a very positive impact on those midlands towns surrounding the great peatfields of the Bog of Allen.[20] But, while keeping people within the country, such developments may have acted to hasten a localized process of migration from the farm to the factory.

Although the geographic imbalance that emerged in the economic character of the island predated 1921, Partition skewed the fiscal profiles

Fig. 7.4. Percentage of the population employed in agriculture at the county level in (a) 1926, (b) 1936 (Free State only), and (c) 1946/51.

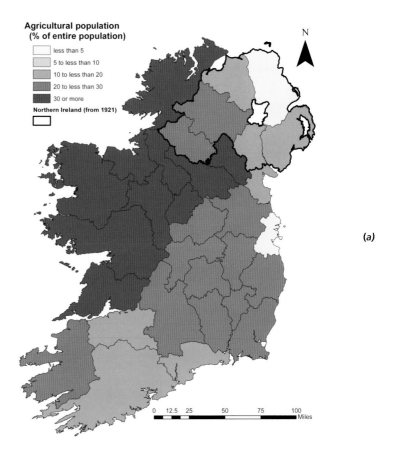

Agricultural population (% of entire population)

- less than 5
- 5 to less than 10
- 10 to less than 20
- 20 to less than 30
- 30 or more

Northern Ireland (from 1921)

(a)

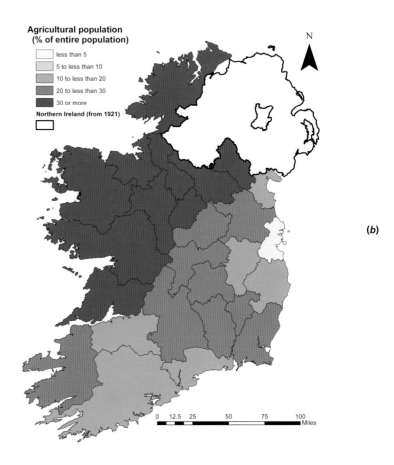

Agricultural population (% of entire population)

- less than 5
- 5 to less than 10
- 10 to less than 20
- 20 to less than 30
- 30 or more

Northern Ireland (from 1921)

(b)

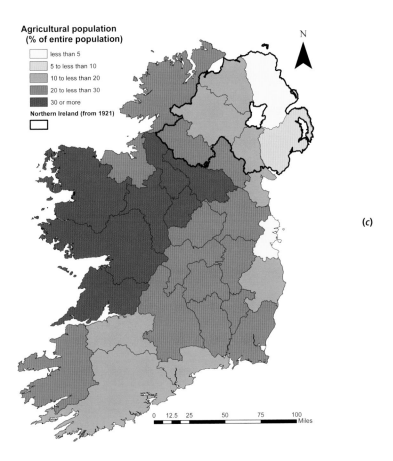

N

(c)

0 12.5 25 50 75 100
Miles

of the two newly created states still further. Johnson and Kennedy have argued that when viewed as a whole, Ireland prior to Partition was not particularly underdeveloped or, indeed, underindustrialized by European standards.[21] This may well be the case, but the political reality was that Ireland could no longer economically be viewed as an integral unit.

Fianna Fáil took power in 1932 and signaled a new and more aggressive economic direction. As premier, Eamon De Valera was determined to use the Irish Free State's financial independence to draw attention to the shortcomings in the state's constitutional freedom. De Valera wanted the removal of the Oath of Allegiance, which recognized the British monarch as titular head of state and which had precipitated the Civil War. When the British refused to renegotiate the treaty settlement, De Valera withheld land annuity payments that had been agreed upon with the London government, which responded by slapping a 20 percent tariff on all Irish agricultural imports.[22] The Irish were thus compelled to respond with a countertax of five shillings per ton on British coal. This policy was undertaken during a drive to industrialize a country with virtually no accessible, high-quality coal seams of its own and was thus self-defeating.[23] This was the start of the "Economic War," which was to have mixed results. On the negative side, the plan was conceptually flawed by the fact that, as previously noted, the Free State's economic fortunes were directly tied to those of the U.K. The economic war altered little in that relationship, and after fiscal hostilities had ceased in the late 1930s, Ireland was still exporting 91 percent of its goods to the U.K. and about twenty-six thousand per annum

Fig. 7.5. Percentage of the population employed in manufacturing at the county level in (*a*) 1926, (*b*) 1936 (Free State only), and (*c*) 1946/51.

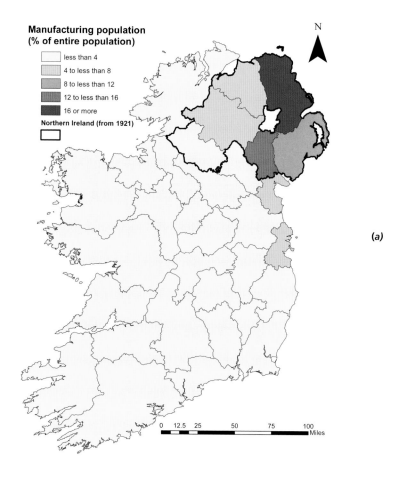

Manufacturing population
(% of entire population)

less than 4
4 to less than 8
8 to less than 12
12 to less than 16
16 or more
Northern Ireland (from 1921)

(*a*)

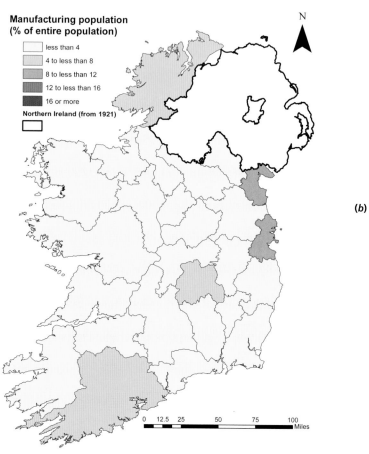

Manufacturing population
(% of entire population)

less than 4
4 to less than 8
8 to less than 12
12 to less than 16
16 or more
Northern Ireland (from 1921)

(*b*)

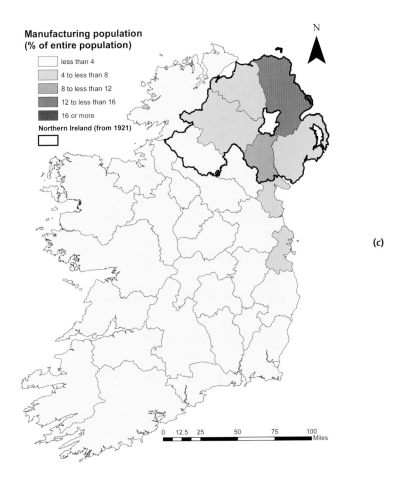

(c)

of its population, mostly to the same place.[24] Thus the population drain continued.

On the positive side, the Economic War was an attempt to free independent Ireland from the trading stranglehold that Britain held over it. The building up of tariff barriers gave indigenous industry a space to thrive, as the emphasis was now placed on the idea of import substitution through increased self-sufficiency.[25] Figure 7.5 shows that this may have had some effect, as industrial employment rose in the mid-1930s Free State, albeit from a very low base. Between 1931 and 1938 industrial employment rose from 110,558 to 166,513, and output increased by 40 percent in the five years to 1936.[26] This is also reinforced in figure 7.6, which shows a substantial decline in the percentage of economically inactive males in the Irish Free State in 1936 in comparison to a decade earlier. It is evident from the map for 1946 that this drive could not be sustained, although the extent to which this was a consequence of World War II is difficult to ascertain.

Northern Ireland, despite its vaunted industrial base, was not without economic concerns in this period. The creation of two separate fiscal entities with differing degrees of autonomy created different problems in both jurisdictions. If the trouble in the south was a surfeit of autonomy and a paucity of judgment, in Northern Ireland the issue has been defined by Buckland as one of an excess of "responsibility without real power."[27] Ulster's decision-making capabilities were severely limited by the terms of the Government of Ireland Act and its peripheral geographical status within the U.K. This, combined with the need to maintain parity with the rest of

Fig. 7.6. Percentage of the male population
economically inactive at the county level in (*a*)
1926, (*b*) 1936 (Free State only), and (*c*) 1946/51.

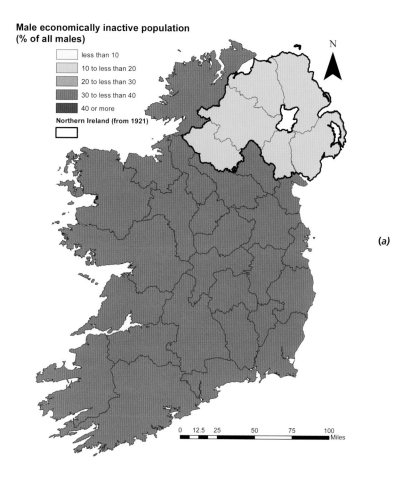

**Male economically inactive population
(% of all males)**

less than 10
10 to less than 20
20 to less than 30
30 to less than 40
40 or more

Northern Ireland (from 1921)

0 12.5 25   50   75   100
Miles

(*a*)

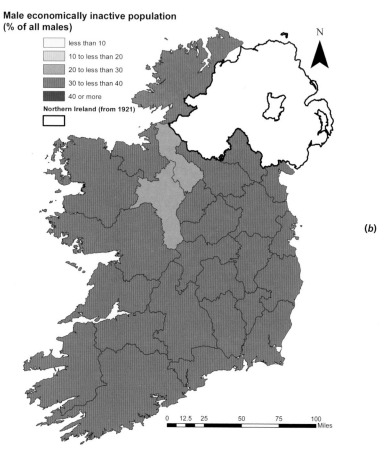

**Male economically inactive population
(% of all males)**

less than 10
10 to less than 20
20 to less than 30
30 to less than 40
40 or more

Northern Ireland (from 1921)

0 12.5 25   50   75   100
Miles

(*b*)

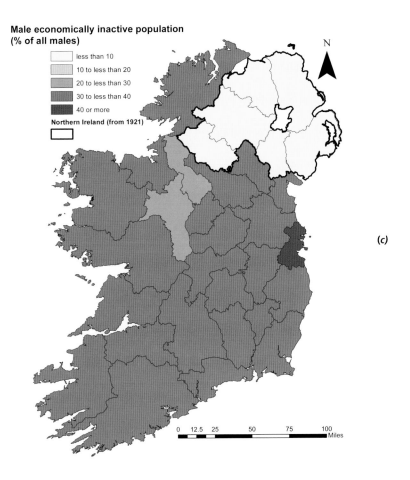

N

(c)

0   12.5   25   50   75   100
Miles

the union in social service provision, placed an unbearable fiscal burden
on the new state, and a pattern of dependence upon the core for economic
subvention would emerge that would characterize Northern Ireland's economic relationship with the rest of the U.K. in the remainder of the twentieth century.[28]

Up to this point northeast Ulster's distinctive industrial character had
always been represented as the basis of its economic good fortune. In the
1930s that industrial character became Ulster's undoing. The depression
hit Northern Ireland extremely hard, harder than any other region of the
U.K., with over one hundred thousand people being out of work in 1935.[29]
The depression was protracted as well as severe. Even in 1938 29.5 percent
of insured industrial workers were still unemployed, compared with 23.8 for
Wales, the worst affected part of Britain.[30] It is unfortunate that data on occupation and employment are unavailable for Northern Ireland in the 1937
census, as we lose the statistical ability to assess this critical period in the
state's economic development. These questions were dropped in order to
facilitate a small Northern Ireland enumeration in 1937 that would then be
synchronized with a larger census to coincide with that planned for Britain
in 1941.[31] The Second World War put paid to that plan, and the next census was not held in Northern Ireland until 1951, by which time a radically
altered economic picture existed. The aggressive economic policies of the
Fianna Fáil government had exacerbated the crisis by cutting Northern
Ireland off from its natural markets for manufactured goods in the south.[32]
Despite all the irredentist rhetoric of the south, embodied in the territorial

claim in articles 2 and 3 of the new 1937 Irish Constitution, the economic policies of De Valera's government had served only to reinforce Partition and inflame sectarian tensions in Northern Ireland.[33]

The Free State's neutrality during World War II was another cause of division on the island, although from a military and political point of view De Valera had little choice other than to adopt an unaligned position.[34] At an economic level the war benefited both states, as it revitalized Northern Ireland's industrial base, which was turned over to the war effort, while the south's agricultural market in the U.K. surged due to wartime food shortages.[35] Underlying the map for 1946/51 in figure 7.5 is the fact that World War II and postwar rebuilding had hauled Northern Ireland's heavy industries out of the dismal and prolonged slump of the 1930s. Also evident from the map is that the war and postwar rebuilding had the opposite effect on the south, reinforcing its reliance on agriculture not only through increased demand but also through the fact that the prewar policy of modest industrialization had increased rather than decreased the south's reliance on external commodities like raw materials.[36] Finally, the south's neutrality meant that it failed to share as fully as Northern Ireland in the fruits of postwar rebuilding programs.

A new period of economic stagnation ensued. During the war emigration naturally declined, but 175,000 people still left the Republic of Ireland area to work in the U.K.'s factories between 1941 and 1946.[37] The immediate postwar period was an economic and demographic disaster for the south, with 500,000 people leaving between 1945 and 1960, the vast bulk due to economic necessity.[38] The lack of economic opportunities and the comparison between north and south could not be more starkly made than in the last map of figure 7.6, which shows the percentage of economically inactive males at the county level in 1946 for the south and 1951 for the north.

Religion and State Building

Figure 7.7 suggests a remarkable consistency in the balance between the three main religions in Ireland in the period between 1911 and 1961, and at the level of the entire island this is indeed accurate. Yet Partition and the conflict that precipitated it set in train religious demographic dynamics that altered Ireland's religious geographies in the mid-twentieth century. Ireland was effectively divided on confessional lines, and religion would inevitably continue to play a powerful role in the construction of identity or identities in both states. Partition did not solve the problem of Ireland's troubled religious geographies, but it did change them. It created two quite different issues of minority management for both the Irish Free State and Northern Ireland. How they responded to these challenges would define both states in the mid-twentieth century.

Partition created an overwhelmingly Catholic population south of the border and a small Protestant minority. By 1926 the 7.4 percent of the Irish Free State's population that were Protestant were economically stratified, geographically scattered, politically immobilized, and, for the most part, left bewildered by their fate under the new dispensation.[39] This is strikingly evidenced by the Church of Ireland's dispatch of a delegation to meet Collins in 1922; the members of this delegation asked him if he would like all

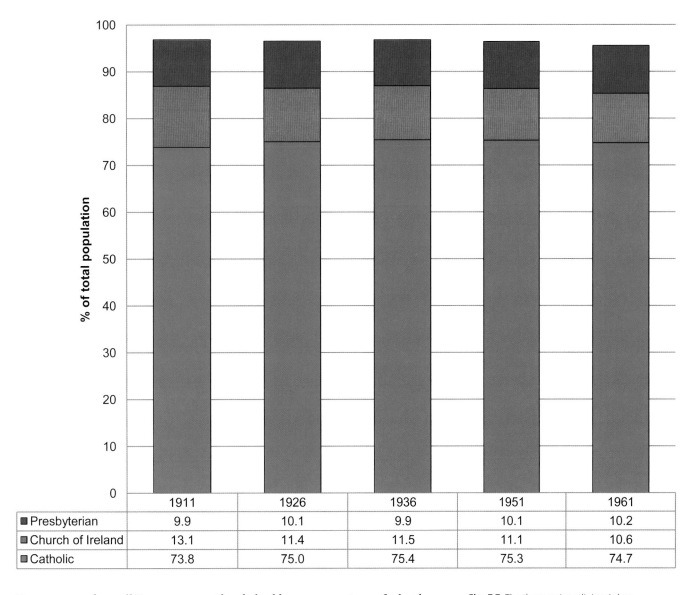

	1911	1926	1936	1951	1961
■ Presbyterian	9.9	10.1	9.9	10.1	10.2
■ Church of Ireland	13.1	11.4	11.5	11.1	10.6
■ Catholic	73.8	75.0	75.4	75.3	74.7

Protestants to leave.[40] Protestants evidently had low expectations of what lay ahead of them under the new regime, although to some it did not appear to be a matter of overwhelming concern. According to the portrait of decaying Anglo-Irish "big house" culture presented in Elizabeth Bowen's novel *The Last September*, "Mrs. Perkins could not promise that there would be much for dinner, but after all this was Ireland and it would be rather jolly."[41] The Anglo-Irish may have been pleasantly surprised to find that the new government lavishly displayed its tolerance toward the Protestant minority with weighted representation in the upper house of the legislature and special funding for Church of Ireland and other minority schools.[42] Protestants also continued to play a disproportionately powerful role in Irish business, commerce, and the professions, accounting for 28 percent of farmers with land over two hundred acres, 38 percent of lawyers, and a remarkable 53 percent of bank officials.[43]

However accommodating the new Free State appeared to be, it continued to become increasingly Catholic, both demographically and culturally. Having dropped by 34 percent over the Partition period, the non-Catholic population of the south fell by a further 49 percent between 1926 and 1961 compared to the Catholic population, which declined by a mere

Fig. 7.7. The three major religions' share of the total population of the island of Ireland between 1911 and 1961.

Fig. 7.8. Catholic populations in (a) 1926, (b) 1936/37, and (c) 1946/51. Earlier maps in this series are available in figure 6.9.

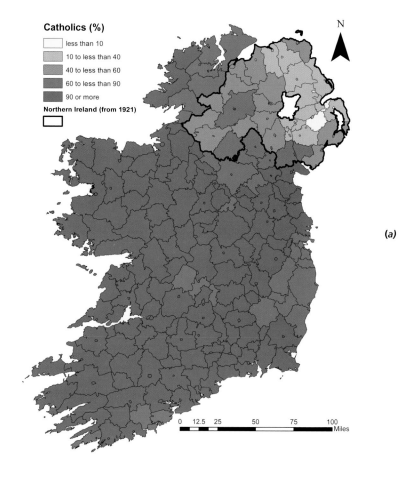

Catholics (%)

- less than 10
- 10 to less than 40
- 40 to less than 60
- 60 to less than 90
- 90 or more

Northern Ireland (from 1921)

N

(a)

0 12.5 25   50   75   100
                            Miles

5 percent over the entire period. This left almost 95 percent of the population being Catholic in 1961, compared to just under 90 percent in 1911. As figure 7.8 shows, the decline of non-Catholics continued to occur in those areas where they had traditionally constituted significant minorities: west Cork, the Pale, and parts of the midlands. Figures 7.9 and 7.10 break the patterns down into their two dominant constituent groups: the Presbyterian Church in Ireland and the Church of Ireland, respectively. As has previously been established, Presbyterians never accounted for a significant part of the population in the three southern provinces of Connacht, Leinster, and Munster, but there were concentrations of them in the Ulster counties left behind by Partition, namely, Cavan, Donegal, and Monaghan. The Church of Ireland population, by contrast, was scattered throughout the southern provinces, and it is in this group that the substantial decline occurs between 1926 and 1946/51.

As table 7.1 shows, the numerical decline of the non-Catholic population was reflected in its becoming more geographically integrated into the population as a whole. The table uses two simple ways of measuring segregation. The *dissimilarity index* measures the proportion of the population that would have to move to make the two groups concerned, in this case Catholics and non-Catholics, evenly distributed. The *isolation index*, known as P*, indicates the chance of a person of one group interacting with a member of the same group if that person is selected at random from the population. If, for example, 80 percent of the population is Catholic, there would be an 80 percent chance of this person being Catholic if Catholics

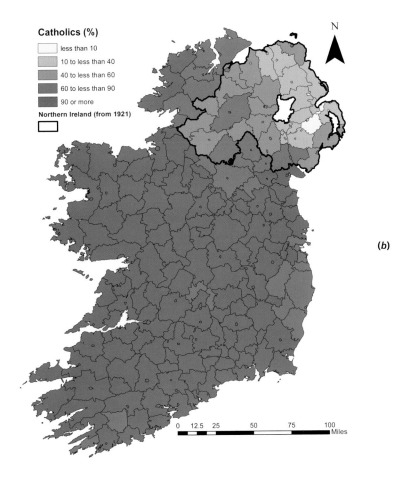

(b)

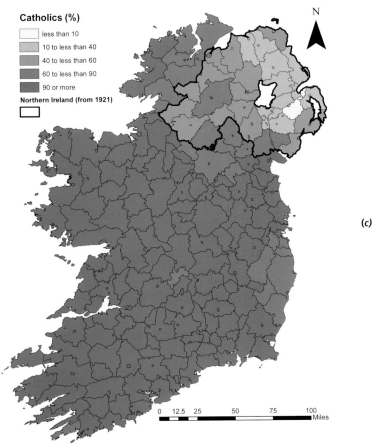

(c)

Fig. 7.9. Presbyterian populations in (*a*) 1936/37,
(*b*) 1946/51, and (*c*) 1961. Earlier maps in
this series are available in figure 6.10b.

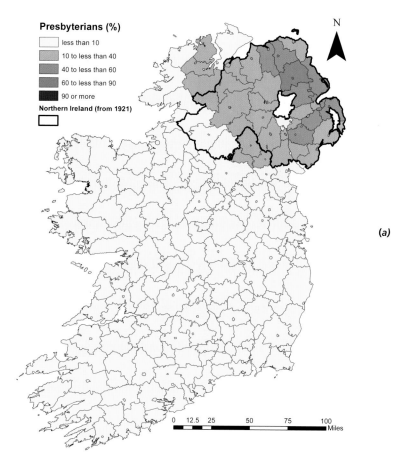

(*a*)

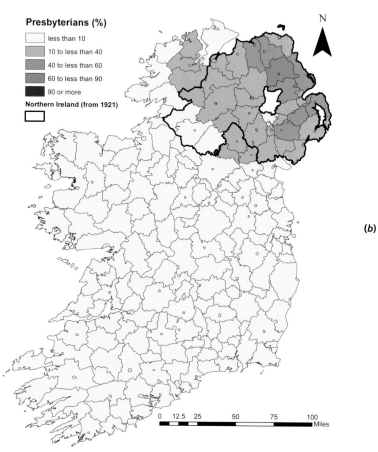

(*b*)

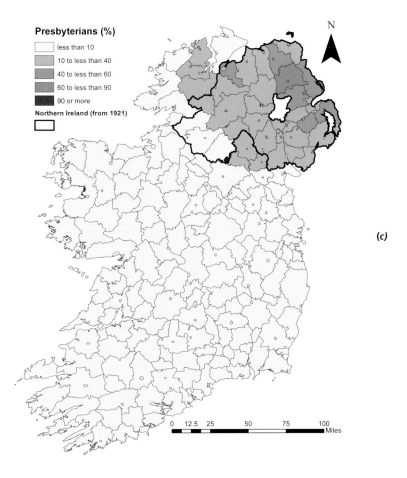

Presbyterians (%)

less than 10

10 to less than 40

40 to less than 60

60 to less than 90

90 or more

Northern Ireland (from 1921)

N

(c)

0   12.5   25   50   75   100
Miles

were evenly distributed across Ireland. As the actual value of P* rises above this expected figure, it indicates that the group is isolated from interacting with members of the other group. In 1911 in the area that was soon to become the Free State, the index of dissimilarity between Catholics and non-Catholics at the urban and rural district scale stood at 39.9 percent. This barely changed over the Partition period; however, from here it fell continuously to reach 34.9 percent in 1961. Similarly, the isolation index shows that non-Catholics were becoming less isolated (this started over the Partition period), while the Catholic population increasingly reflected the overall population distribution.

At a legislative level, increasing censorship and a prohibition on divorce in the mid- to late 1920s were viewed by many Protestants as curbing individual freedoms and marginalizing their faith,[44] especially as these restrictions were occurring at a time when Episcopalians were liberalizing their attitudes toward contraception.[45] The Catholic Church was increasingly willing to intervene in political matters where it felt that policy would detrimentally affect the spiritual welfare of its flock.[46] A good example of this was the "mother and child scheme" of 1951, which aimed to provide non-means-tested, free healthcare for all mothers and children up to the age of sixteen. The bill received political support, but when clerical opposition to it became known, the politicians backed down. The church's fears centered around the possibility of Catholic mothers being treated by doctors who were not of their religion and of the scheme being used as a Trojan horse for the introduction of contraception or abortion.[47]

Fig. 7.10. Church of Ireland populations in (*a*)
1936/37, (*b*) 1946/51, and (*c*) 1961. Earlier maps
in this series are available in figure 6.10a.

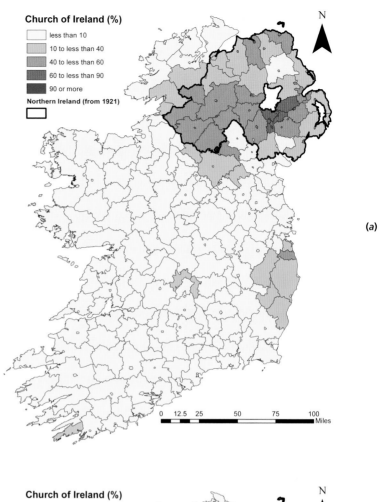

(*a*)

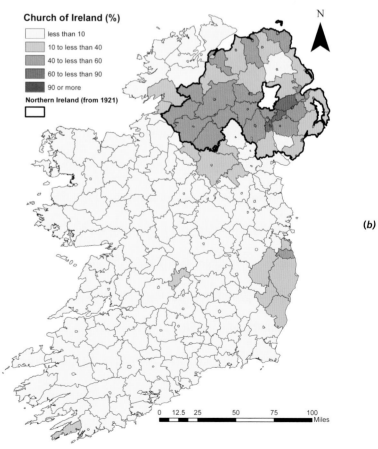

(*b*)

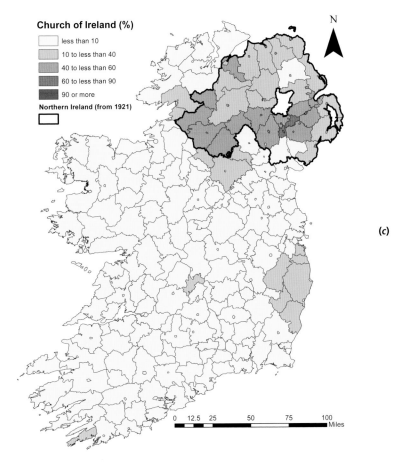

Church of Ireland (%)

- less than 10
- 10 to less than 40
- 40 to less than 60
- 60 to less than 90
- 90 or more

Northern Ireland (from 1921)

N

(c)

0   12.5   25   50   75   100
                                         Miles

To critics in Northern Ireland, the increasing Catholicization of the southern state was epitomized in De Valera's 1937 Constitution, which recognized the "special position" of the Catholic Church. This was much less than Rome actually wanted, which was a text that recognized "the one, true Church,"[48] but the reality of Irish society in 1937 was that the Catholic Church did occupy a "special position," and any attempt to ignore that would have been disingenuous. The constitution also recognized the other denominations present in Ireland, including the Jewish congregation at a time when, in the European political context, such recognition made a bold statement. The constitution was a pragmatic and dignified recognition of the role of religion in Irish society while maintaining the de jure separation of church and state by its refusal to endow any creed.[49]

In Northern Ireland the picture was somewhat different. Partition had bulwarked a sustainable Protestant majority, but the authorities north of the

Table 7.1. Indexes of segregation for the Free State/Republic, 1911–1961

	1911	1926	1936/37	1946/51	1961
Index of dissimilarity	39.9	39.7	38.6	37.8	34.9
P* Catholics	90.4	93.1	93.8	94.6	95.1
Expected P* Catholics	89.3	92.6	93.4	94.3	94.9
Difference	1.1	0.5	0.4	0.3	0.2
P* non-Catholics	19.7	13.9	12.1	10.8	9.5
Expected P* non-Catholics	10.7	7.4	6.6	5.7	5.1
Difference	9.0	6.4	5.6	5.1	4.4

Note: Figures have been calculated using data interpolated onto 1961 urban and rural districts.

Fig. 7.11. The Catholic population of Northern Ireland in (*a*) 1926, (*b*) 1937, (*c*) 1951, and (*d*) 1961.

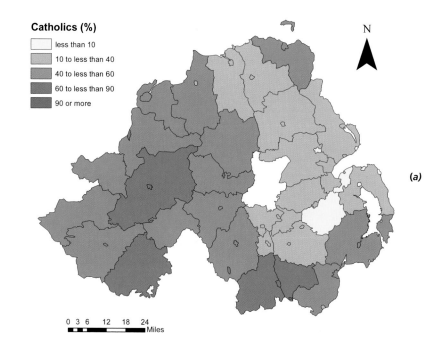

(*a*)

Catholics (%)

- less than 10
- 10 to less than 40
- 40 to less than 60
- 60 to less than 90
- 90 or more

0 3 6   12   18   24
Miles

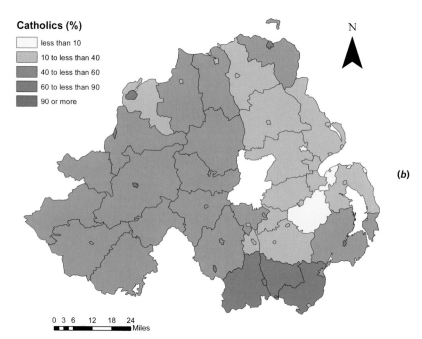

(*b*)

Catholics (%)

- less than 10
- 10 to less than 40
- 40 to less than 60
- 60 to less than 90
- 90 or more

0 3 6   12   18   24
Miles

border were left with a large minority population that remained between 33 and 35 percent of the population between 1911 and 1961. The distribution of these populations is shown in the time series of maps of Northern Ireland in figures 7.11, 7.12, and 7.13. Figure 7.11 shows that even after the Partition settlement there was a substantial area of the new jurisdiction (in fact, most of the area west of Lough Neagh and the River Bann) in which Catholics were the majority. In addition to these areas, within the "heartland" of Presbyterianism was A. T. Q. Stewart's ancient "bridgehead" of Scots Catholicism in the Glens of Antrim, not to mention a burgeoning and resentful urban working class in the Pound Loney, New Lodge, and

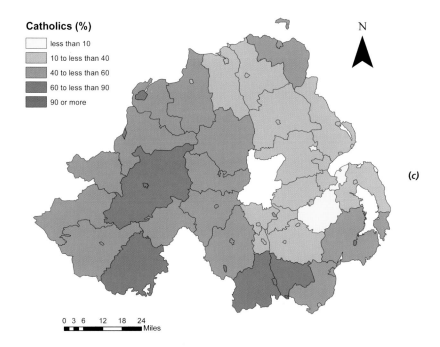

N

(c)

0 3 6 12 18 24
Miles

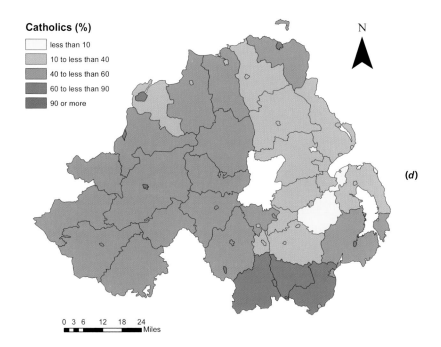

N

(d)

0 3 6 12 18 24
Miles

Ardoyne districts of Belfast. Table 7.2 shows that in some ways Northern Ireland does not seem to have been much more segregated geographically than the south, the only major difference between these values and those shown in table 7.1 being that Catholics and non-Catholics seem almost equally isolated from each other.

Compounding the demography (or, rather, interrelated to it) was the knowledge that very many of these Catholics did not recognize the very legitimacy of the existence of the state. This stood in stark contrast to the position in the south, where the much smaller Protestant minority largely accepted the new dispensation, throwing their lot in with the protreaty

Fig. 7.12. The Presbyterian population of Northern Ireland in (*a*) 1926, (*b*) 1937, (*c*) 1951, and (*d*) 1961.

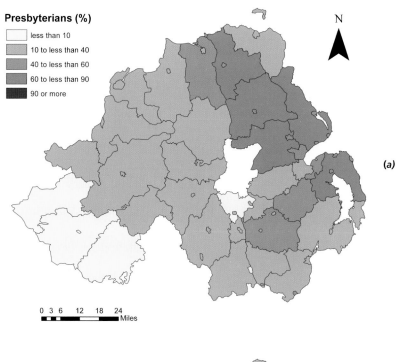

(*a*)

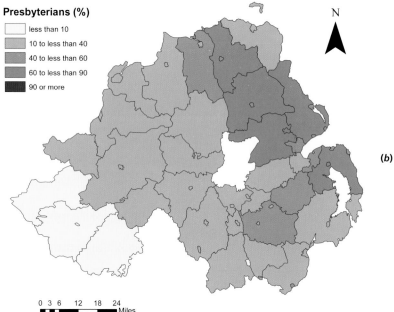

(*b*)

Cumann na nGaedhael Party as the best defender of their interests.[50] This "disloyalty" of Catholics in the north cast them as "natural enemies" of the state in which they lived.[51]

How did the Northern Ireland government respond to this dilemma? It had two options. The first was to conciliate Catholics and attempt to win them over to the new regime; the second was to marginalize them still further from the political system. The government chose the latter. Through the abolition of proportional representation and the gerrymandering of constituency boundaries to artificially maintain unionist majorities—most infamously, to keep control of the city of Derry—the government further

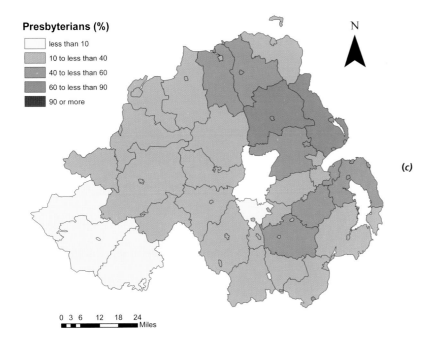

Presbyterians (%)

- less than 10
- 10 to less than 40
- 40 to less than 60
- 60 to less than 90
- 90 or more

N

(c)

0 3 6   12   18   24
Miles

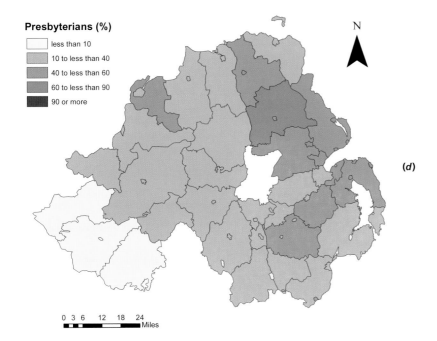

Presbyterians (%)

- less than 10
- 10 to less than 40
- 40 to less than 60
- 60 to less than 90
- 90 or more

N

(d)

0 3 6   12   18   24
Miles

undermined any faith the Catholic population had in the institutions and exercise of power in the new state.[52] Any lingering doubts could be finally dispelled when the prime minister, James Craig, spoke in 1934 of a "Protestant parliament for a Protestant state," which is what had been achieved.[53] During the entire period of devolved government in Northern Ireland, the greatest legislative achievement of the Nationalist Party, which represented the aspirations of most Catholics, was the less than controversial Wild Birds Act of 1931.[54]

Catholics could also take some responsibility for Northern Ireland's failings, as they largely opted out from the start. It was little surprise that the

Fig. 7.13. The Church of Ireland population of Northern Ireland in (*a*) 1926, (*b*) 1937, (*c*) 1951, and (*d*) 1961.

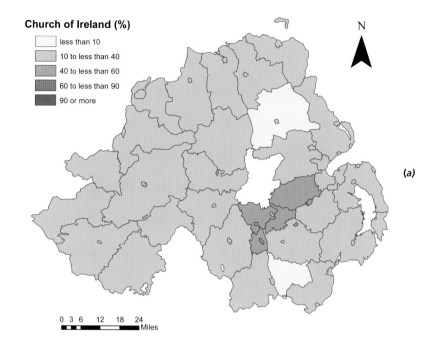

Church of Ireland (%)

- less than 10
- 10 to less than 40
- 40 to less than 60
- 60 to less than 90
- 90 or more

N

(*a*)

0 3 6 12 18 24 Miles

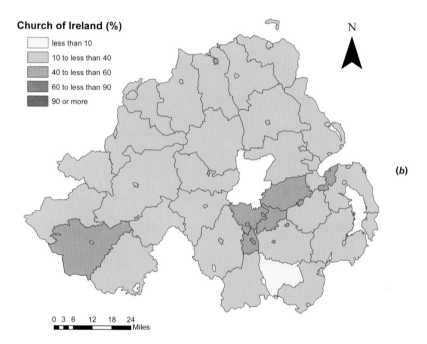

Church of Ireland (%)

- less than 10
- 10 to less than 40
- 40 to less than 60
- 60 to less than 90
- 90 or more

N

(*b*)

0 3 6 12 18 24 Miles

majority population would gladly take the decisions for Catholics. There were few examples of genuine attempts to move beyond the religious divide in Northern Ireland, but the one remarkable attempt came early on with Lord Londonderry's educational reforms of the early 1920s. He was determined to establish a religiously integrated system of education at the elementary level, a plan that gained the support of MPs who wished to avoid "sectarianism and denominationalism in schools."[55] Opinion differs on who was to blame for sabotaging the scheme first: the Catholic Church and its teachers refused to cooperate with the reform committee, while Protestant clergymen actively mobilized against it.[56] One thing is certain:

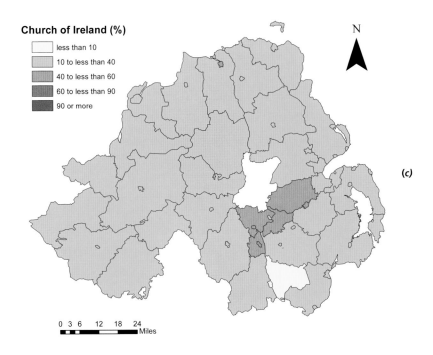

Church of Ireland (%)

- less than 10
- 10 to less than 40
- 40 to less than 60
- 60 to less than 90
- 90 or more

N

(c)

0 3 6 12 18 24 Miles

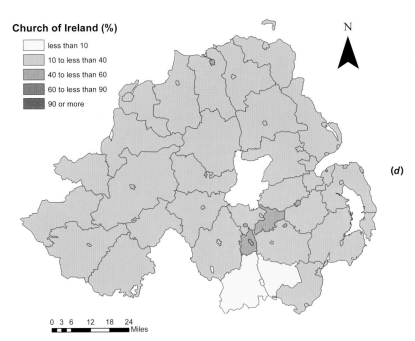

Church of Ireland (%)

- less than 10
- 10 to less than 40
- 40 to less than 60
- 60 to less than 90
- 90 or more

N

(d)

0 3 6 12 18 24 Miles

Table 7.2. Indexes of segregation for Northern Ireland, 1911–1961

	1911	1926	1936/37	1946/51	1961
Index of dissimilarity	35.2	36.7	35.9	35.5	34.3
P* Catholics	43.7	43.0	42.9	43.4	44.0
Expected P* Catholics	34.5	33.5	33.5	34.4	34.9
Difference	9.2	9.6	9.5	9.0	9.1
P* non-Catholics	70.3	73.4	72.6	74.3	72.1
Expected P* non-Catholics	65.5	66.5	66.5	65.6	65.1
Differences	4.8	6.8	6.1	8.7	7.0

Note: Figures have been calculated using data interpolated onto 1961 urban and rural districts.

the plan did not happen, and for a time at least, Catholics and Protestants were united in their determination to keep apart.[57] Had things worked out differently at the beginning, events a generation or two later might have taken a different course. In this sense the state really was "maimed . . . at the start."[58]

Conclusions

Partition did not create the divisions in Ireland's religion, society, and economy, it merely formalized existing rifts. In the period between the 1920s and 1950s the trends that divided Ireland continued, while the political gulf between the two states increasingly widened. This occurred not only as a result of policy decisions but indirectly through the creation of two competing political cultures in which the necessity for accommodation and toleration had been removed by the spatioreligious territorial settlement of 1921 and confirmed in the Boundary Commission report of 1925, which created an overwhelmingly Catholic south and a Protestant-dominated north. Figure 7.14 shows that over the longer period from 1911 to 1961 the process of religious realignment had continued, with the proportion of the Protestant population in areas south of the border continuing to fall. This was not balanced by a discernible decline in the Catholic proportions north of the border. On the contrary, over this period the Catholic proportion of the population in the greater Belfast area, specifically in a transect west of the city, increased considerably.

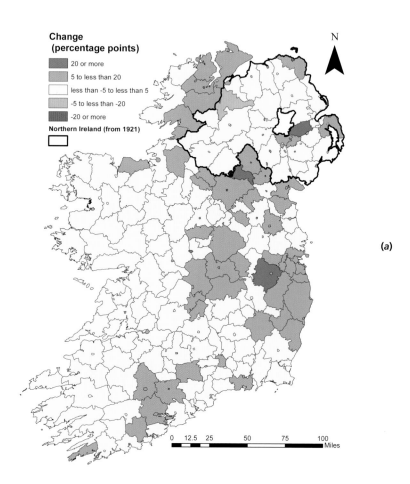

Fig. 7.14. Percentage point change between 1911 and 1961 in the proportion of the populations of (a) Catholics, (b) Church of Ireland, and (c) Presbyterians.

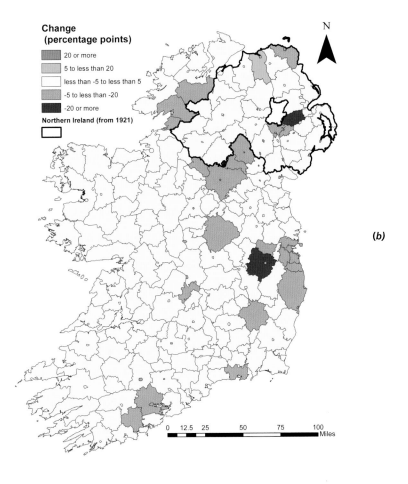

(b)

0   12.5   25   50   75   100
                                         Miles

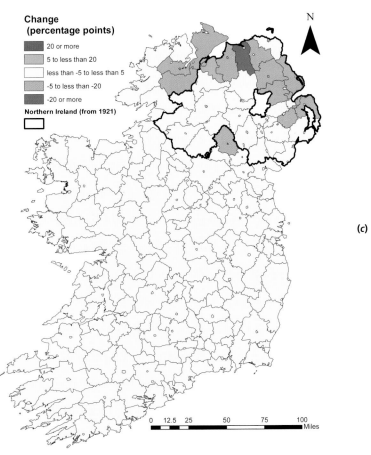

(c)

0   12.5   25   50   75   100
                                         Miles

In 1948 Taoiseach John A. Costello surprised not only Britain and Northern Ireland but the population of the south as well when he declared Ireland was to become a Republic. What was most telling about this episode was that it became emblematic of how far the two parts of the island had strayed from political realities, even since Partition. Costello had earlier spoken of the hopes of uniting Ireland, while the declaration only made this less likely.[59] The northern prime minister promptly called an election, which acted as a referendum to firmly endorse the territorial settlement.[60] Division was not only an external theme but one that defined relations between different religious groups within both states in this period. The religious homogeneity that Partition imposed on the Irish Free State became a wider feature of its political and social development, whereas the relative hazardousness of the northern Protestant position fostered insecurity and obduracy. In the coming decades both states would be forced to confront the legacy of the resentments that had piled up in this period.

Toward the Celtic Tiger: The Republic, 1961 to 2002

8

Up to this point the story of the south of Ireland's economic fortunes has been characterized by an agricultural economy blighted by stagnation and failure. From the beginning of the 1960s a series of policy changes would occur in the Republic that would have profound consequences for the state not simply in the economic sphere but in the social, political, demographic, and even religious realms. It may, at first glance, be tempting to view the period from 1961 to 2002 in terms of a linear path toward economic and social maturity, but such a simplistic teleological interpretation bears little resemblance to what was an extremely turbulent period in the state's short history.

A New Departure: Lemass, Whitaker, and *Economic Development*

The 1950s in the Republic of Ireland has come to be seen as a time of economic and social stagnation. By that time people could reflect on the bitter reality that over the thirty years since independence, the state had failed in its primary obligation—to provide the economic means for people to remain living in their own country. De Valera's 1943 dream of a land of cozy homesteads was instead a place of empty homesteads, much of the "sturdy youth" having departed for New York, London, or Manchester.[1] More than four hundred thousand people left the south of Ireland between 1951 and 1961, most of them because of economic necessity.[2] Yet it was the sense of failure that crystallized in this decade that led to a renewed determination to resolve the Republic's ongoing population crisis.[3]

During the 1950s the southern polity matured and began to develop the means to reflect on its own flawed progress through the publication of a variety of indigenous journals sharing ideas in the intellectual and administrative spheres.[4] A document published at the end of the decade by a senior civil servant in the Department of Finance entitled *Economic Development* was the result of this fermenting of reflective practice with the negative motivations of shame and inadequacy at the central administration's impotence in the face of massive ongoing emigration. T. K. Whitaker's report formed the basis for Taoiseach Seán Lemass's First Programme for Economic Expansion. The aim was to drive up industrialization and employment through incentive schemes to encourage foreign direct investment (FDI) as well as to stimulate indigenous entrepreneurial growth. In 1952 the Industrial Development Authority (IDA) was given responsibility for attracting foreign capital, a function it would fulfill with considerable success.[5] It was recognized that the policies of protectionism and import substitution since independence had made Irish industry uncompetitive and inward-looking.[6]

The First and Second Programmes for Economic Expansion proved to be remarkably successful, at least in the short term. An economic growth rate of 4 percent was achieved before the 1960s were out, and foreign firms were employing a quarter of the industrial workforce by 1977. Critically, the Programmes broadened Ireland's economic horizons by breaking the dominance of the asymmetric trading relationship with the U.K., with two-thirds of exports now going beyond the British market. The success of the schemes is demonstrated by the profound and lasting effect that they had on the Republic of Ireland's demography, with the population growing by one hundred thousand during the 1960s.[7] Unfortunately, the boom of the 1960s would not be part of a continuous upward economic trend. The oil crisis of 1974 exposed Ireland's vulnerability to foreign markets and the danger of an overreliance on FDI as the prime motor of economic development.[8] Nevertheless, in demographic terms a corner had been turned, despite the economic collapse of the early 1980s, which brought the state to the brink of bankruptcy and saw a fresh wave of young and better-qualified migrants leave the country.[9] So *Economic Development* was not the start of a continuous progression toward the success seemingly achieved in the Celtic Tiger era, but it altered the Republic of Ireland so fundamentally and in so many ways in the period since the 1950s that its significance must be acknowledged at the outset of this exploration of the data.

Population: Staunching the Flow

Economic Development embodied a sort of missionary zeal that did not stem from a desire to create wealth as an end in itself; instead, it sought increased employment as the means by which independent Ireland's population problem could be remedied.[10] As figure 8.1 shows, the reopening of Ireland's economy to the rest of the world managed to effect a positive demographic change. However, while this trend reversed the pattern of overall population loss from the Republic, the economic renaissance also dramatically altered the distribution of its population. Preexisting trends of urban expansion and rural decline were rapidly accelerated, with growth being almost a completely urban phenomenon; indeed, almost all growth since 1961 has occurred in the towns and cities. The rural population has continued to stagnate throughout the period, with only slight evidence of tentative growth between the last full censuses.

Figure 8.2 maps population density by rural and urban districts for main census dates between 1971 and 2002. Maps for earlier years are shown in figure 7.2. The maps show that population densities continued to drop across a very wide part of the country even after the overall demographic decline had been stalled. The growth of suburbanization, particularly in the greater Dublin area, is also apparent, as is population growth around the regional centers of Cork, Limerick, and Galway. Thus, while the national trend may have been reversed, the trend of urban growth and rural decline with the accompanying shift in population from the west to the east evident since the Famine has, in fact, continued. Therefore, over this period the Republic has become increasingly divided between a depopulated rural periphery and a rapidly growing urban and suburban core.

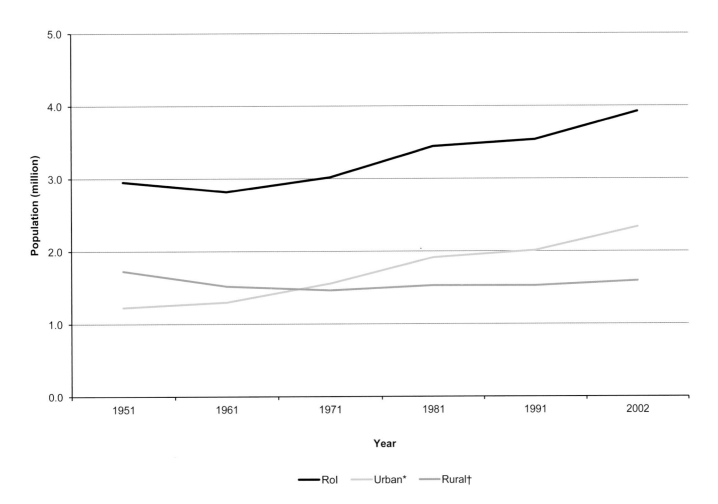

Legend: ——Rol ——Urban* ——Rural†

Figure 8.3 shows just how dramatically the transition has occurred over the period from 1951 to 2002 and, more specifically, how demographically divided the state has become between a developing east coast and a declining western seaboard. Decline is a feature of all peripheral regions but has been most extreme in the northwest of the Republic, in Counties Leitrim and Roscommon and the east part of Mayo. This contrasts with the unrelenting growth of the greater Dublin area—Counties Dublin, Kildare, and Wicklow—clearly evinced in figure 8.4 and table 8.1. This graph and accompanying table show how Dublin and its surrounding counties have grown to account for almost 40 percent of the Republic's population over the course of the half-century since 1951.

Thus, what emerges is a portrait of a country undergoing urbanization at a very rapid rate. In one sense, the Republic of Ireland was little different from any other western European country that experienced an industrial revolution. What made the Irish experience special was the speed with which it was transformed from an essentially preindustrial society through industrialization and on to postindustrialization. In countries such as Britain these developments evolved over centuries, while in southern Ireland the entire process occurred within a matter of decades. The same degree of dynamism is evident in the course of the Republic's urban development during the same time. Returning to figure 8.3, we can see that the Borough of Dublin actually experienced little change in its population during the second half of the century. In contrast, districts surrounding the

Fig. 8.1. Republic of Ireland rural, urban, and total populations between 1951 and 2002. * Defined as all urban areas with a population of or in excess of 1,500. † Defined as including urban areas with a population below 1,500.

Fig. 8.2. Population density at rural and urban district levels for the Republic of Ireland in (*a*) 1971, (*b*) 1981, (*c*) 1991, and (*d*) 2002. Earlier maps from this series are shown in figure 7.2.

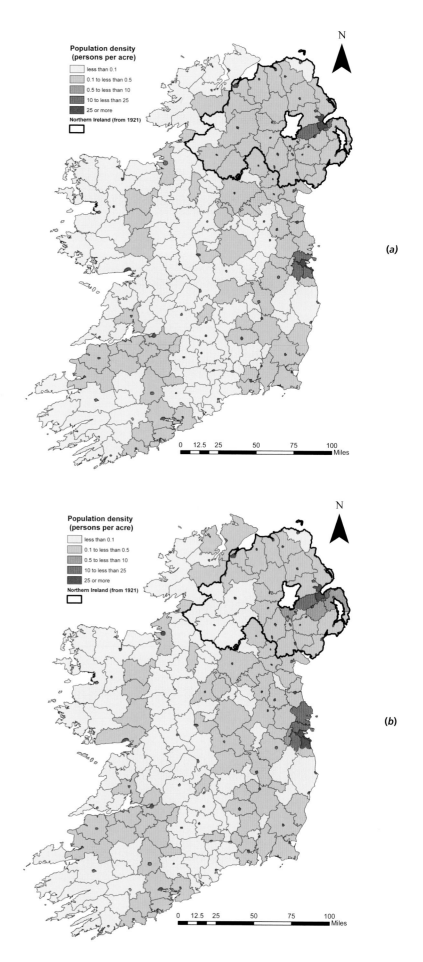

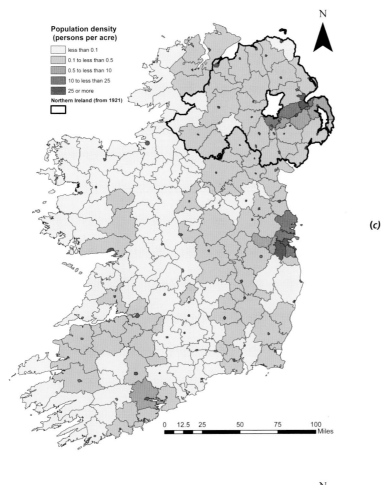

**Population density
(persons per acre)**

less than 0.1
0.1 to less than 0.5
0.5 to less than 10
10 to less than 25
25 or more
Northern Ireland (from 1921)

(c)

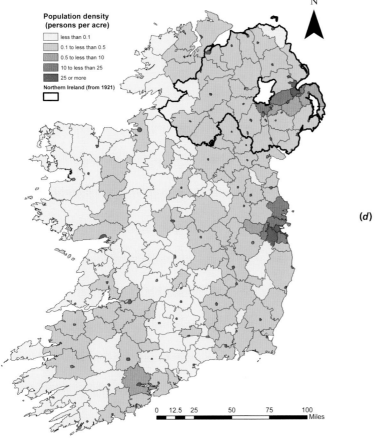

**Population density
(persons per acre)**

less than 0.1
0.1 to less than 0.5
0.5 to less than 10
10 to less than 25
25 or more
Northern Ireland (from 1921)

(d)

The Republic, 1961 to 2002   141

Fig. 8.3. Percentage of population change at rural and urban district levels for the Republic of Ireland between 1951 and 2002.

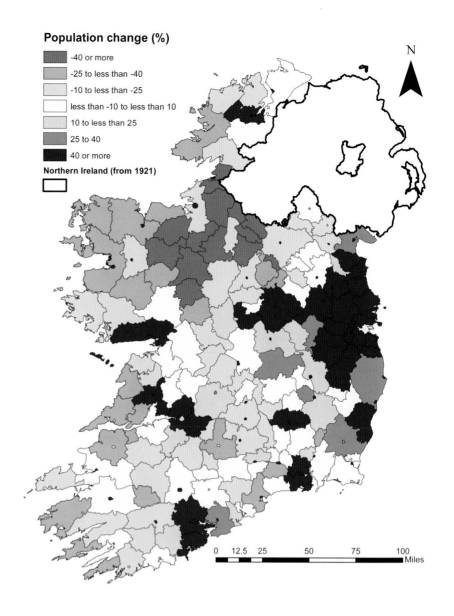

Population change (%)

- -40 or more
- -25 to less than -40
- -10 to less than -25
- less than -10 to less than 10
- 10 to less than 25
- 25 to 40
- 40 or more

Northern Ireland (from 1921)

```
0   12.5   25        50        75       100
                                        Miles
```

city experienced massive growth, with Balrothery and Dublin South seeing growth rates in excess of 500 percent. In large part this was due to the fact that the city became effectively saturated in demographic terms, leading to vast suburban development from the mid-1960s.

Employment was identified as the first priority in reversing the downward population trend; however, once that problem appeared to have been successfully addressed, housing the growing populace emerged as an urgent issue. As part of Whitaker's recommendations, spending on housing and other social infrastructure had initially been cut back as part of the First Programme for Economic Expansion. This policy changed after the shortfall led to a crisis exemplified by the collapse of a block of tenements in Dublin's inner city that killed four people and underlined the decrepit state of much of the existing and heavily overcrowded residential stock.[11] The answer was the building of large peripheral housing estates. In general these were low-density schemes, but the Republic also experimented with the high-rise planning initiatives then in vogue across Europe. The huge Ballymun project on the northern edge of Dublin, which was capable of housing some twenty thousand souls, fared much like its counterparts

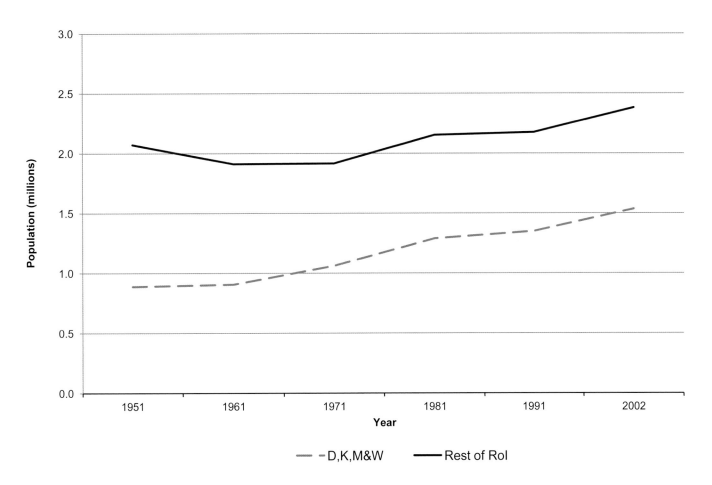

— - D,K,M&W ——— Rest of RoI

Table 8.1. Table of population growth in Counties Dublin, Kildare, Meath, and Wicklow relative to the rest of the Republic of Ireland between 1951 and 2002

Fig. 8.4. Population growth in Counties Dublin, Kildare, Meath, and Wicklow (D, K, M, and W) relative to the rest of the Republic of Ireland between 1951 and 2002.

	Dublin, Kildare, Meath, and Wicklow	RoI area	Rest of RoI	% Dublin, Kildare, Meath, and Wicklow
1951	888,386	2,960,593	2,072,207	30.0
1961	906,347	2,818,341	1,911,994	32.2
1971	1,062,067	2,978,248	1,916,181	35.7
1981	1,290,154	3,443,405	2,153,251	37.5
1991	1,350,595	3,525,719	2,175,124	38.3
2002	1,535,446	3,917,203	2,381,757	39.2

across the continent and soon became a byword for crime, deprivation, and metropolitan decline.[12] While other estates provided families with better quality housing, the rapid building of vast suburbs in places like Tallaght to the southwest of Dublin was poorly planned, with little thought given to infrastructural and social provision.[13] A lack of schools, shops, and other facilities meant that in the early decades these areas were akin to dormitory settlements with little identity of their own, as residents had nowhere or no reason to interact with each other. As a result, people continued to look back toward the city for practical and emotional reasons.[14]

These problems can at least be regarded as evidence of demographic vitality as the cities struggled to accommodate their rapidly growing populations, but this vitality was far from evenly distributed. Figures 8.1 through 8.4 show how growth in the Republic could be classified largely as an urban and eastern phenomenon. Population decline in rural areas was seen as

Table 8.2. Table of population growth of towns of 500 to fewer than 1,500 inhabitants between 1951 and 2002

	Towns < 1,500	RoI	Towns < 1,500 as %
1951	97,397	2,954,352	3.3
1961	131,122	2,817,869	4.7
1971	129,348	3,018,566	4.3
1981	136,354	3,443,403	4.0
1991	174,180	3,536,311	4.9
2002	184,062	3,926,005	4.7

both a metaphor for and a contributory factor to a deepening social crisis. Nancy Scheper-Hughes's controversial anthropological study of a west Kerry village identified particularly high female emigration rates as one contributory factor in the prevalence of mental illness amongst bachelor men left behind.[15] The picture in the small market towns of the west did not appear much more positive either. John Healy's 1968 *"Nobody Shouted Stop": The Death of an Irish Town* examined the decline of his native Charlestown in east Mayo, while the same sense of drift permeates John Waters's *Jiving at the Crossroads*, which focused on his hometown of Castlerea in neighboring County Roscommon during the economic collapse of the early 1980s.[16] At first glance the statistics do not appear to support this sense of pessimism. Table 8.2 shows the cumulative population of towns of between five hundred and fewer than fifteen hundred inhabitants between 1951 and 2002. Over this period this class of settlement, to which both Charlestown and Castlerea belonged, increased their proportional contribution to the population of the Republic. Patterns diverge when we compare these two towns with two settlements of comparable size in 1951, both within fifteen miles of central Dublin. Figure 8.5 illustrates how urban growth itself was heavily skewed toward the greater Dublin area as tiny villages such as Celbridge mushroomed into large satellite towns while the towns in the west stagnated.

Economic Shift: From Stagnation to the Celtic Tiger

Despite the reversals and setbacks that characterized the gestation of the Republic's economic rise, the state's population was beginning to grow and move rapidly. Driving these seismic changes in Irish demography was the engine of industrialization.

Figure 8.6 shows the changing pattern of manufacturing employment precipitated by the Programmes for Economic Expansion. This sequence of maps makes two very significant points. First, they show that the industrial drive from the late 1950s succeeded in pushing up employment in the secondary sector, and, more importantly, that drive managed to do so across a wide swath of the state. In the past, Ireland's industrial development had been highly concentrated in the northeast of the island and to a lesser extent in Dublin. Now an attempt was being made to disperse the development of productive employment more equitably. This was the spatial expression of what Lemass, quoting John F. Kennedy, dubbed the "rising tide that would lift all boats."[17] An Foras Tionscail, the Underdeveloped

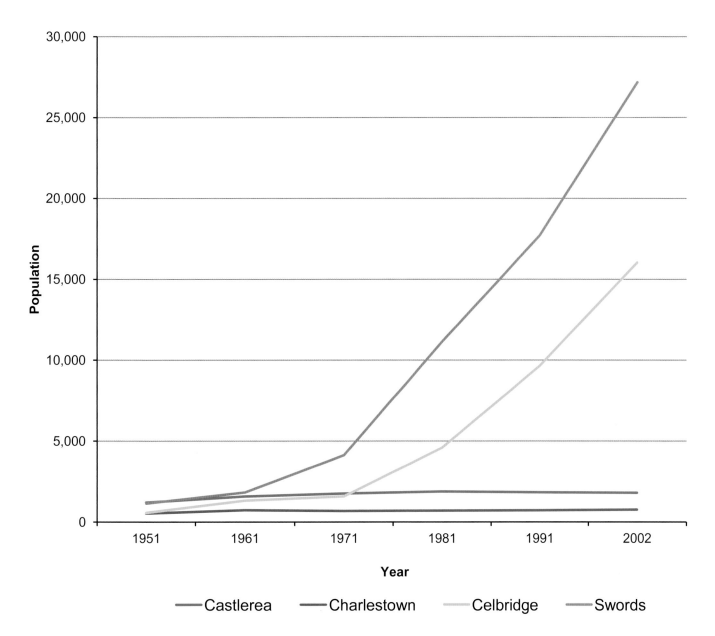

Areas Board, was charged with ensuring that the benefits of industrialization were spread to the poorest parts of the country.[18]

Along with incentivization schemes, infrastructural improvements also helped to attract foreign companies to the country. The establishment of the Shannon Free Enterprise Zone next to Shannon Airport in County Clare best embodied both the practical and the ideological elements of the new economic strategy. This was a large industrial estate with special tax concessions for companies operating within its confines.[19]

The Buchanan Report of 1968 recommended that the benefits of industrial development could be maximized by the relative concentration of future factories in a number of regional centers.[20] However, for several reasons the government instead opted for a policy of maximum dispersal.[21] The moral argument for such a decision was that the government wished to do as much as possible to prevent the flow of people from the countryside, even if the alternative was intraregional rather than international movement; however, the other motivation was perhaps more base and reflected

Fig. 8.5. Population change in four towns between 1951 and 2002: Castlerea (Roscommon), Charlestown (Mayo), Celbridge (Kildare), and Swords (Dublin).

Fig. 8.6. Employment in the manufacturing sector as a percentage of the adult population at the county level for the Republic of Ireland in (a) 1961, (b) 1981, and (c) 2002. Earlier maps from this series are available in figure 7.5.

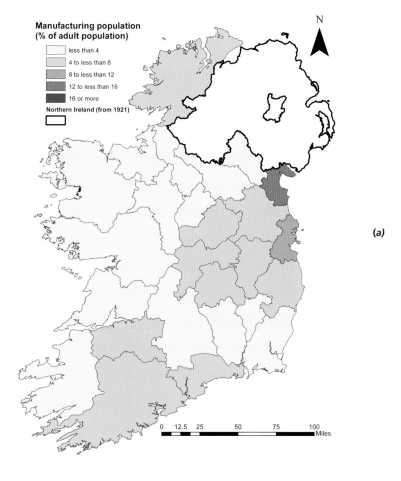

(a)

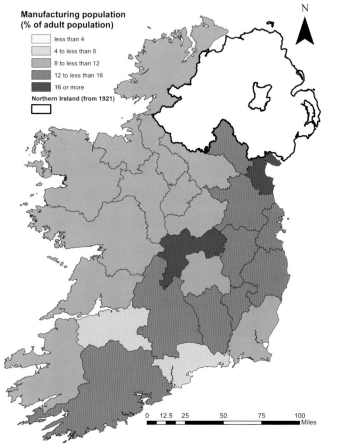

(b)

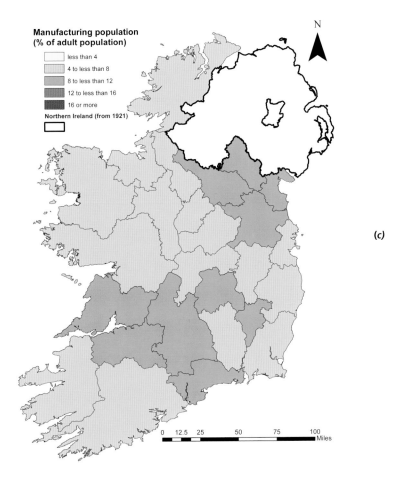

N

(c)

0   12.5   25   50   75   100
                                          Miles

more about the evolution of Irish political culture. A lack of substantive ideological division between the two major parties in the state, Fianna Fáil and Fine Gael, meant that personality has historically been more significant than policy in determining differences between the two and that a cult of personality has often emerged around more charismatic leaders.[22] A culture of political clientelism existed in which the individual voter was allied to the party more by personal loyalty and patronage than by deeper ideological beliefs.[23] It was perhaps an exaggerated characterization, traditionally associated more with rural areas than with the towns and cities, but the prevalence of the model should not be underestimated. There is a long and, in some senses, democratic tradition of direct special pleading between the voter and politicians operating at the highest levels of government, and this has meant that rural areas have historically had a political influence exceeding their demographic weight.[24] It does not follow that this led to a sort of solidarity between rural communities. In reality, it fostered an ugly tradition of egotism, with small towns squabbling over the distribution of political favors, and this has characterized debates over industrial location in independent Ireland since the 1930s.[25] In fairness, motivations were not completely parochial—the neglect of the state's railway network meant that even the shortest commutes between small towns and provincial centers, let alone from more rural areas, were completely impractical, and this was another significant factor militating against the development of a provincial strategy.[26]

The dispersal policy led to the building of factories in villages and rural locations. This was in itself a mixed blessing. Those locations now had a substantial source of local employment, meaning that young people no longer had to leave in search of work. On the negative side, the policy created an economic culture of dependence upon an often exogenous company and plugged the most rural communities into the erratic globalizing trade cycle. When many of these companies folded during the recession of the early 1980s, these communities were left effectively high and dry, with no alternative source of employment and with residents hampered by a very specific skill set tailored to a particular industry.[27]

This leads to the second crucial point that arises from figure 8.6. The pattern of manufacturing employment in the Republic of Ireland in the second half of the twentieth century shows how quickly the state moved from being an almost preindustrial society to a postindustrial one. Manufacturing employment peaked in 1981 and subsequently began to decline. In large part this was the result of the protracted recession of the 1980s, demonstrated in figure 8.7, which shows unemployment by county between 1961 and 2002. The highest levels of unemployment are shown in 1991, and the stubbornly high levels in the Republic's most isolated county, Donegal, are evidence of the rural character of the problem.

The collapse in manufacturing was in part offset by the emergence of other sectors of the economy. Figures 8.8 and 8.9 show the increase in people working in the clerical and the commercial/financial areas of the economy, respectively, since the 1970s. The financial sector is concentrated on the east coast, but the number of people working in offices increased across the state during that period. Underlining the fundamental shifts that have occurred in the Republic's economy is figure 8.10, which clearly identifies the rapid decline in agricultural employment at every census since 1961. Taken as a whole, these four series of maps show starkly how the country moved through the model of economic progression from a largely agrarian primary stage in 1961 to a diversified postindustrial tertiary economy within the space of forty years. Despite this, in 2002 Dublin and the east coast still had the majority of the clerical, commercial, and financial jobs, while the agricultural sector was largest in the west and the midlands.[28]

The expansion of access to third-level education, as higher education is known in Ireland, was central to the Republic's economic fortunes in the latter half of the twentieth century. The 1970s saw a substantial increase in the number of places available in third-level education as the number of universities and colleges increased.[29] Much of this expansion came against a background of intense economic trauma during the early 1980s, when the state came close to bankruptcy and Taoiseach Charles Haughey warned that "as a community we are living way beyond our means."[30] The result was much closer governmental control over university research activities with a view to presenting a strong economic justification for the investment. This led to an emphasis on subjects such as business studies, information technology, and applied sciences.[31] Figure 8.11 shows that the number of people with college degrees in the Republic increased rapidly between 1971 and 2002. The availability of an educated young workforce in the emerging new technology sector combined with a highly successful tradition of industrial incentivization created the firm foundations for the

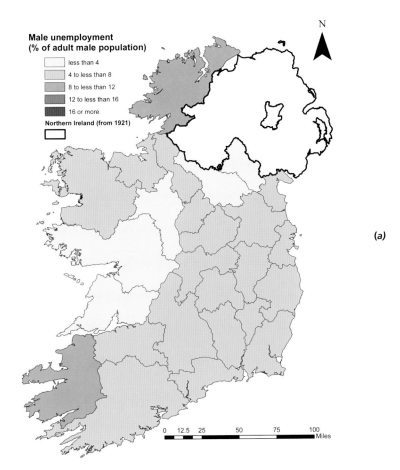

(a)

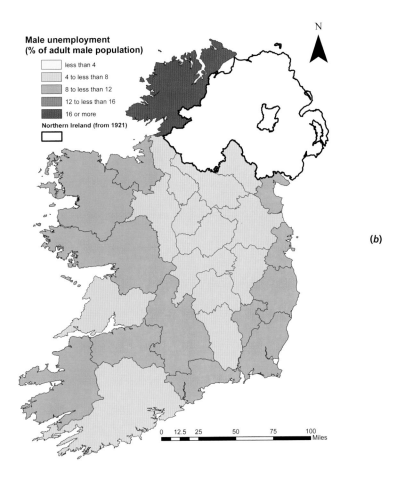

(b)

Fig. 8.7a and 8.7b. Male unemployment at the county level for the Republic of Ireland in (a) 1971, (b) 1981. Earlier maps from this series are available in figure 7.6. (*Continued on the next page*)

The Republic, 1961 to 2002

Fig. 8.7c and 8.7d. Male unemployment at the county level for the Republic of Ireland in (*c*) 1991, and (*d*) 2002. Earlier maps from this series are available in figure 7.6.

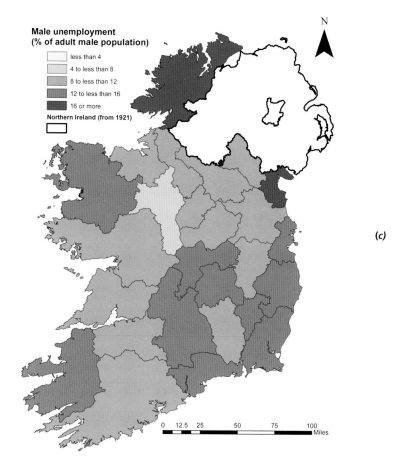

(*c*)

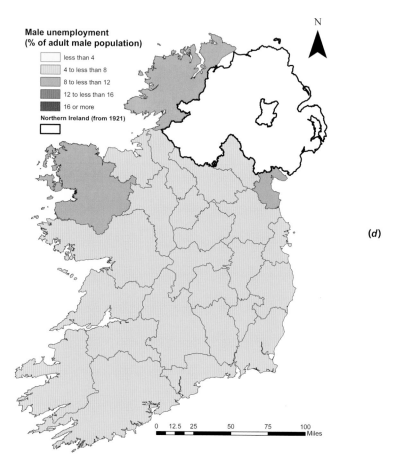

(*d*)

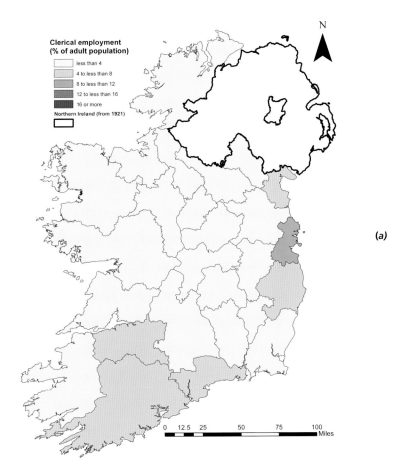

(a)

Fig. 8.8a and 8.8b. Population employed in the clerical sector as a percentage of the adult population at the county level for the Republic of Ireland in (*a*) 1971, (*b*) 1981. (*Continued on the next page*)

Clerical employment
(% of adult population)

less than 4
4 to less than 8
8 to less than 12
12 to less than 16
16 or more
Northern Ireland (from 1921)

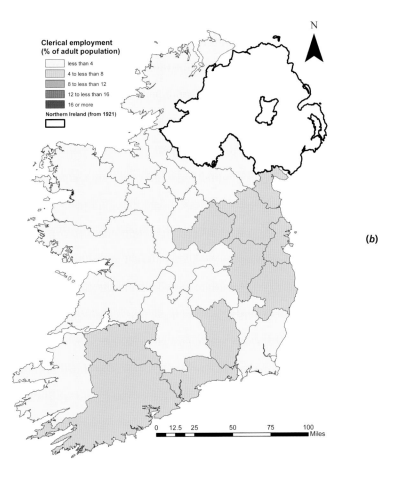

(b)

Clerical employment
(% of adult population)

less than 4
4 to less than 8
8 to less than 12
12 to less than 16
16 or more
Northern Ireland (from 1921)

Fig. 8.8c and 8.8d. Population employed in the clerical sector as a percentage of the adult population at the county level for the Republic of Ireland in (c) 1991, and (d) 2002.

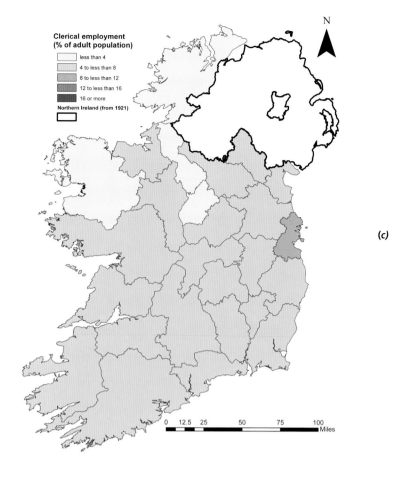

(c)

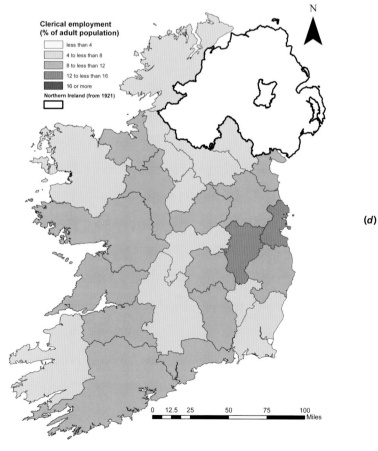

(d)

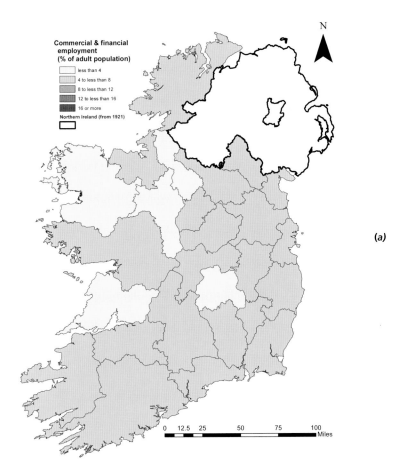

(a)

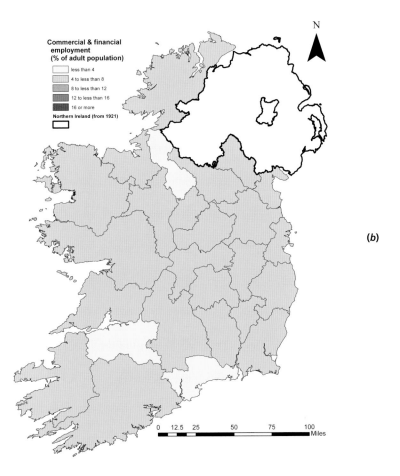

(b)

Fig. 8.9a and 8.9b. Population employed in the commercial and financial sectors as a percentage of the adult population at the county level for the Republic of Ireland in (a) 1971, (b) 1981. (*Continued on the next page*)

Fig. 8.9c and 8.9d. Population employed in the commercial and financial sectors as a percentage of the adult population at the county level for the Republic of Ireland in (c) 1991, and (d) 2002.

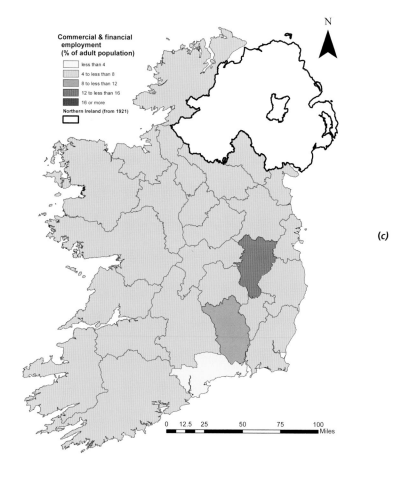

(c)

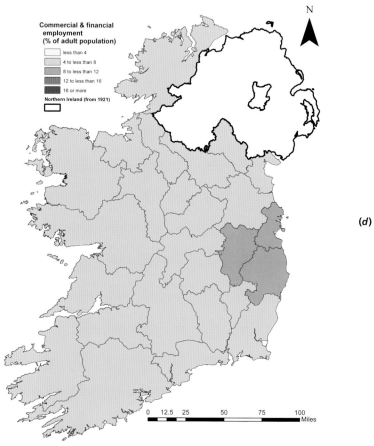

(d)

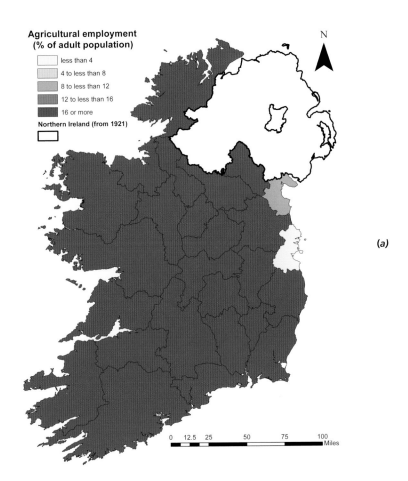

**Agricultural employment
(% of adult population)**

- less than 4
- 4 to less than 8
- 8 to less than 12
- 12 to less than 16
- 16 or more

Northern Ireland (from 1921)

N

(a)

0 12.5 25 50 75 100
Miles

Fig. 8.10a and 8.10b. Population employed in agriculture as a percentage of the adult population at the county level for the Republic of Ireland in (*a*) 1961, (*b*) 1981. Earlier maps from this series are available in Figure 7.6. (*Continued on the next page*)

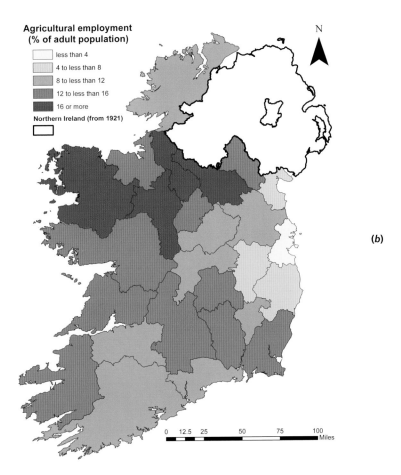

**Agricultural employment
(% of adult population)**

- less than 4
- 4 to less than 8
- 8 to less than 12
- 12 to less than 16
- 16 or more

Northern Ireland (from 1921)

N

(b)

0 12.5 25 50 75 100
Miles

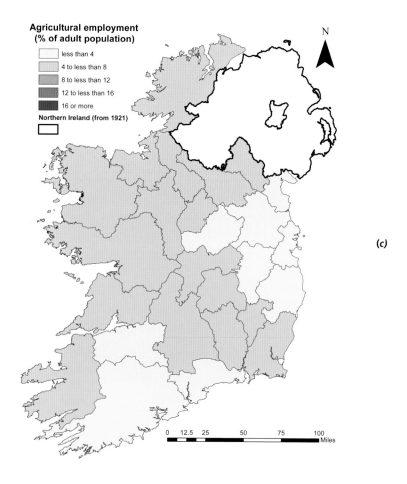

Fig. 8.10c. Population employed in agriculture as a percentage of the adult population at the county level for the Republic of Ireland in (c) 2002.

Agricultural employment (% of adult population)

- less than 4
- 4 to less than 8
- 8 to less than 12
- 12 to less than 16
- 16 or more

Northern Ireland (from 1921)

N

(c)

0   12.5   25   50   75   100
Miles

rapid economic growth of the "Celtic Tiger" period. It was these conditions, as well as excellent marketing of the perceived "quality of life" benefits, that prompted multinational corporations such as Microsoft and Intel to set up not only assembly operations but extensive research and development infrastructure in the Republic.[32] By the end of the century the Republic was both the fastest-growing economy and the fourth-richest country in the world. It had truly earned the moniker of the Celtic Tiger.[33]

The Changing Role of Religion in Irish Society

Just as the economy of the south of Ireland entered an entirely new phase in this period, so too did religion. The period after Partition saw the Protestants' share of the south's population decline markedly and their geographical spread narrow, while at the same time the Catholic Church's role in the country and its government increased. Protestantism had declined to such an extent that after 1961 the census no longer went to the trouble of distinguishing between the Church of Ireland and Presbyterians at local levels, simply reporting the total number of Protestants instead. What emerges is a period in which secularism started to rise as people began to either refuse to answer the question on religion or declare no religious affiliation. This transformation has been shaped in part by the process of modernization but to a far greater extent by developments that have affected the position of the organization to which the overwhelming majority of the Republic's population have traditionally subscribed—the Catholic Church.

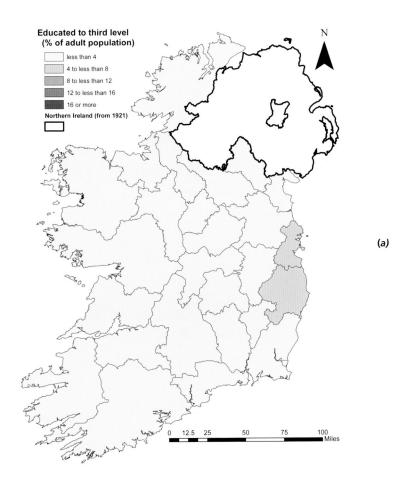

**Educated to third level
(% of adult population)**

- less than 4
- 4 to less than 8
- 8 to less than 12
- 12 to less than 16
- 16 or more

Northern Ireland (from 1921)

N

0 12.5 25   50   75   100
Miles

(a)

Fig. 8.11a and 8.11b. Percentage of college-educated adults by county for the Republic of Ireland in (a) 1971, (b) 1981. (*Continued on the next page*)

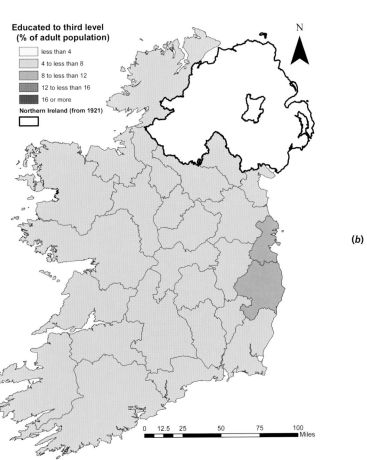

**Educated to third level
(% of adult population)**

- less than 4
- 4 to less than 8
- 8 to less than 12
- 12 to less than 16
- 16 or more

Northern Ireland (from 1921)

N

0 12.5 25   50   75   100
Miles

(b)

Fig. 8.11c and 8.11d. Percentage of college-educated adults by county for the Republic of Ireland in 1981, (c) 1991, and (d) 2002.

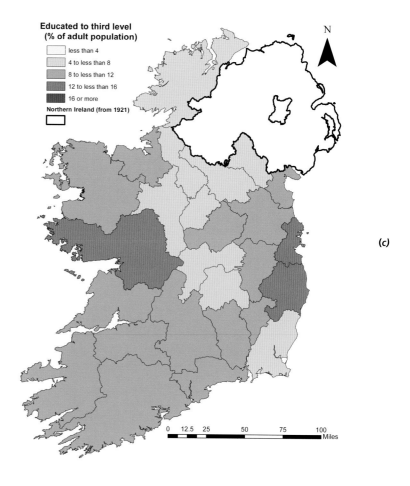

(c)

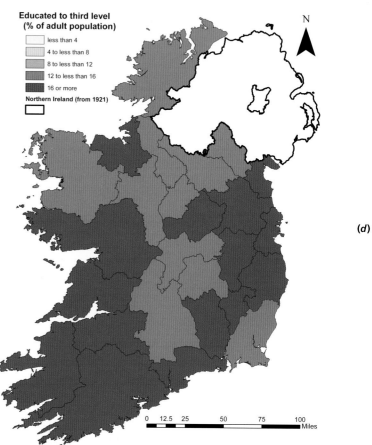

(d)

The series of maps in figure 8.12 illustrates that while Catholics continued to predominate across the state, toward the end of the period significant non-Catholic minorities had emerged or, in the case of west Cork, reemerged. Explaining this trend is not straightforward, but it may in part be understood as a result of a modest process of countercultural migration to the Atlantic seaboard over the past forty years that began to alter the religious geographies of the coastal zone as young Catholic natives have been replaced by an older generation of better-off immigrants from Britain, the Netherlands, Germany, and other predominantly Protestant regions of northern Europe.[34] In more recent years, the Republic's towns and cities have been transformed by immigration as a result of the success of the Celtic Tiger period. The effect of these trends has been to reverse the increasing numerical dominance of Catholics in the south, which peaked in 1961, when Catholics represented 94.9 percent of the population. This had dropped to 91.5 percent in 1991 and then dropped further to 88.4 percent by 2002, the lowest proportion measured by any census. At the same time non-Catholics have become more mixed with Catholics. The index of dissimilarity dropped from 39.4 percent in 1961 to 19.3 percent in 2002. As the proportion of non-Catholics has risen, so has their isolation index, from 9.5 percent in 1961 to 13.5 percent in 2002, but the gap between the actual and expected isolation indexes has fallen from 4.4 percentage points to a mere 2 percentage points, suggesting that this increasing non-Catholic minority is far less segregated geographically than the traditional non-Catholics of earlier periods.

The graph in figure 8.13 shows that there is also evidence that other new religious trends were beginning to emerge. While the Republic remains an overwhelmingly Catholic place, at least as defined by the census, for the two main Protestant denominations this period was particularly significant in that for the first time since the mid-nineteenth century both bodies increased their membership in the twenty-six-county area. Between 1961 and 2002 the Church of Ireland grew by 30 percent, while Presbyterians increased by 56 percent. This compares to the more modest change in the Catholic population, which only rose by 7.3 percent. The resurgence in the numbers of Protestants can be ascribed to the recent influx of immigrants who have begun to alter the religious geographies of the Republic's towns and cities.[35]

Perhaps more striking than the dynamics of change between the major religions has been the increase in the number of Irish people who no longer choose to affiliate with any religion. Figure 8.14 shows how those who identify themselves as of "no religion" or who refuse to supply a response remain relatively small by western European standards, but their numbers have increased rapidly during the latter half of the twentieth century. Just how rapid that increase has been is borne out by figure 8.15, which indicates the intercensal percentage change between those adhering to a particular religion and those either refusing to answer or of no religion. The growth, albeit from an extremely low base, has been extraordinary. Between 1961 and 1971 the number of people identifying themselves as having no religion increased by close to 600 percent, while the growth in the religious population was closer to 6 percent. Over the same decade, the percentage of people refusing to answer the religion enquiry increased by 729 percent.

Fig. 8.12. Catholic populations in (*a*) 1961, (*b*) 1981, and (*c*) 2002. Earlier maps from this series are available in figure 7.8.

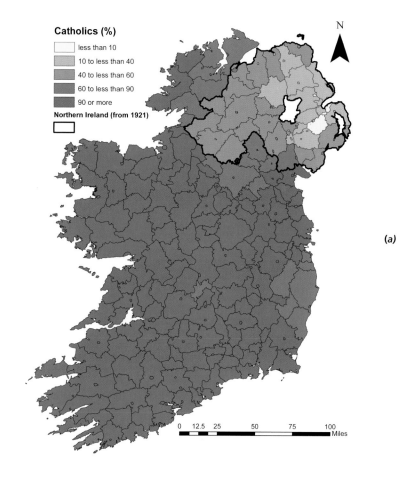

(*a*)

Catholics (%)

less than 10
10 to less than 40
40 to less than 60
60 to less than 90
90 or more

Northern Ireland (from 1921)

0 12.5 25   50   75   100
Miles

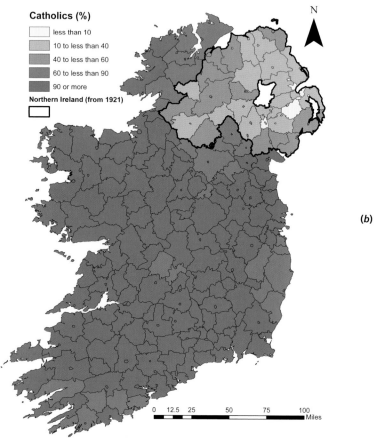

(*b*)

Catholics (%)

less than 10
10 to less than 40
40 to less than 60
60 to less than 90
90 or more

Northern Ireland (from 1921)

0 12.5 25   50   75   100
Miles

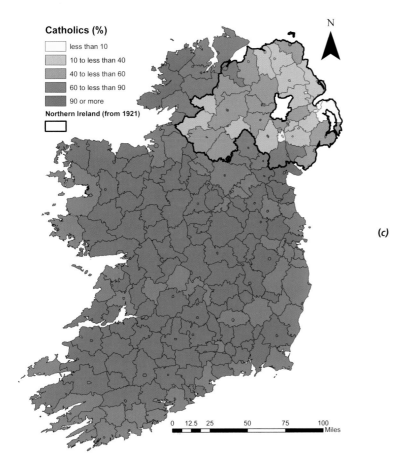

Catholics (%)

- less than 10
- 10 to less than 40
- 40 to less than 60
- 60 to less than 90
- 90 or more

Northern Ireland (from 1921)

N

(c)

0 12.5 25   50   75   100
Miles

Fig. 8.13. Responses to the religion question in Republic of Ireland censuses between 1961 and 2002 categorized by major religious group.

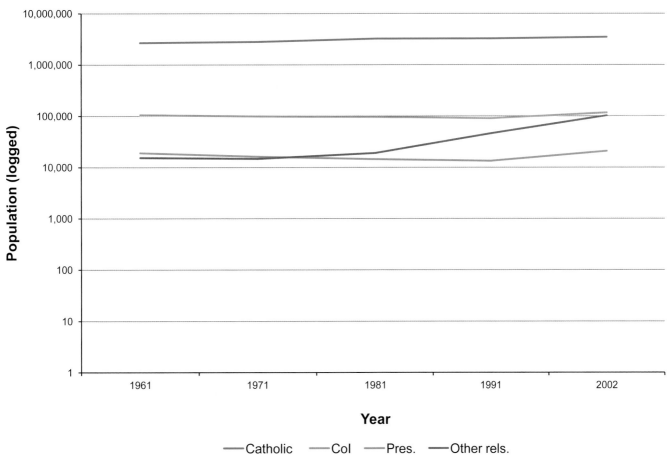

—Catholic —CoI —Pres. —Other rels.

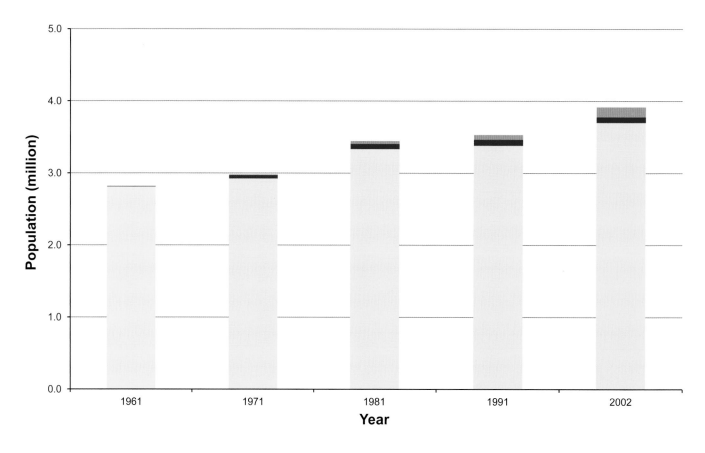

Religious affil.   ■ No info. supplied   ■ No religion

Fig. 8.14. Responses to the religion question in Republic of Ireland censuses between 1961 and 2002 categorized by simple affiliation, nonaffiliation, or refusal to submit information.

In 1961 the number of people who professed to having no religious affiliation accounted for less than 0.04 percent of the Republic's total population. By 2002 this had increased to 3.5 percent, with the numbers more than doubling between 1991 and 2002 from 66,270 to 138,264. Interestingly, the percentage change in those refusing to give information in the period between 1991 and 2002 was negative, perhaps suggesting that people may be more willing to articulate their outright atheism or agnosticism than had previously been the case.

Why have these changes occurred? Part of the explanation may be a response to the Troubles at a philosophical or a pragmatic level. At a philosophical level, during the 1970s in particular people may have expressed their objection to religious labels, which they somehow saw as a measure of complicity in the tribal conflict in Northern Ireland. At a pragmatic level, during the worst periods of violence some respondents may have been fearful about drawing attention to their religion in an area of the Republic (e.g., nearer the border) where communal tensions emanating from the conflict in Northern Ireland may have been greater.[36]

It has been argued that modernization leads to secularization, although many sociologists now dismiss the idea that a decline in religion is a feature of industrializing and industrialized societies as they gradually disengage from folk and religious social systems. Nevertheless, this idea does make temporal sense in the Republic's case, as the onset of secularization has coincided largely with the rapid economic development of the country since the early 1960s and the resulting social dislocation. However, inter-

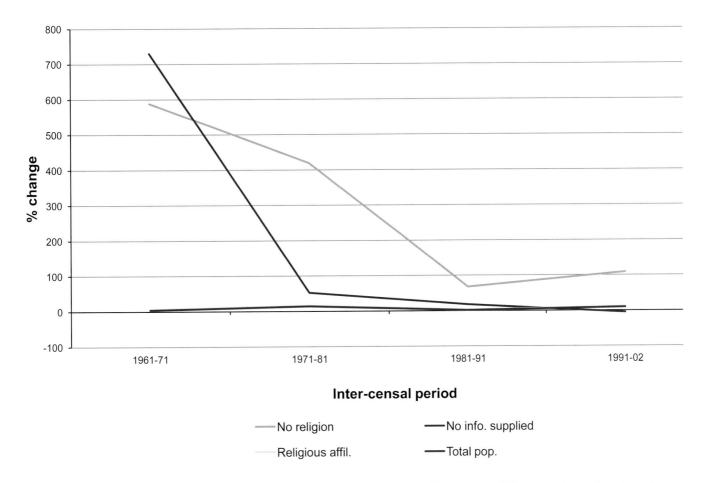

Inter-censal period

No religion   No info. supplied

Religious affil.   Total pop.

nal dynamics are also likely to have been more significant. Writing in 1986, Máire Nic Ghiolla Phádraig concluded that religious structures were still extremely powerful in the Republic and that the vast majority of people still deferred to the Catholic Church on moral issues. Her conclusion, however, is astonishingly prescient. She states that the church is "a vulnerable body and one in which a loss of credibility by the bishops on a single important issue could jeopardize the persistence of commitment among a majority of their flock."[37] In the intervening years this is precisely what has happened, as the church has lost its vaunted position in Irish society in response to a harrowing array of child abuse scandals since the 1990s.[38] This decline in status is perhaps most starkly borne out by the figures for attendance at mass. When Nic Ghiolla Phádraig was writing in the mid-1980s, about 90 percent of the Republic's population was still attending mass on a weekly basis. Twenty years later this had fallen to about 55 percent.[39] Whether this is a consequence of specific disillusionment with the Catholic Church in Ireland or a consequence of broader secularization is hard to determine; however, what it does demonstrate is that the decline in importance in religious identity—as measured by the census—was mirrored if not amplified by a decline in religious practice in the form of church attendance.

Fig. 8.15. Intercensal percentage change in responses to the religion question in Republic of Ireland censuses between 1961 and 2002.

Conclusions

This chapter has explored spatially and statistically some of the remarkable changes that affected the Republic of Ireland during the second half of the twentieth century. Despite the oscillations in the global trade cycle, the scale and speed of those changes across time and space had left the country in a far better economic position in 2002 than it was in the early 1950s. However, the gap between richest and poorest had increased, social inequalities were still entrenched, and the distribution of wealth had remained heavily skewed.[40] At both cultural and social levels, through the homogenizing impacts of free trade and globalization, a more "relaxed view of Irishness" may have been achieved, but the power of "advertising prose and Muzak" in redetermining Irish identity in this period is the era's defining motif.[41] From a religious perspective, the rise in industrialization and wealth has coincided with a decline in the influence of Catholicism numerically, spatially, and morally within society.

Stagnation and Segregation: Northern Ireland, 1971 to 2001

The late twentieth century saw a stark contrast between the experiences of the Republic of Ireland, described in the previous chapter, and those of Northern Ireland over the same period. While the Republic saw rapid economic progress and a decline in religious divisions, the situation in Northern Ireland was almost the reverse. Between 1971 and 2001 Northern Ireland saw rapid economic change as its traditional industries declined. At the same time it experienced a prolonged sectarian conflict in the form of the Troubles, during which more than three thousand people died. The complexity of the situation means that the next three chapters will be devoted to covering Northern Ireland over this period. Chapter 9 looks at demographic, economic, and social change, stressing that in many ways Northern Ireland's experience was typical of declining heavy industrial regions, albeit with a unique spatioreligious undertone. Chapter 10 then moves to exploring the patterns of violence that occurred during the Troubles, which started in the late 1960s and ended with the various ceasefires of the late 1990s. Chapter 11 draws these two threads together, focusing on Belfast, the area in which these themes had their largest impacts.

Recent censuses provide ways to assess geographical change in Northern Ireland at much more local levels than have been possible elsewhere in this book. The 1971 census was the first "modern" census in that data were published in an electronic format. These data are available in the form of the Northern Ireland grid square product,[1] which provides population counts for many social, economic, and demographic variables for square cells with sides of 1 km and 100 m.[2] Uniquely within the U.K. and Ireland, counts for these cells were also provided for 1981, 1991, and 2001, which, because they use the same geographical units in different censuses, allow spatially detailed mapping of trends through time. Maps produced this way are unconventional in that they are not tied to the usual spatial units used for social mapping, such as the rural and urban districts seen earlier in this volume, but instead provide counts for where the population is actually located. This, in turn, allows us to discuss the geographies of Northern Ireland and how they changed over this key period in significantly more detail than is available for the rest of the book.

This chapter begins by considering some selected ways in which Northern Ireland changed between 1971 and 2001, outlining some key economic and demographic themes. It then examines some of the geographical outcomes of these developments using the 1 km Northern Ireland grid square product. Following this, the chapter explores questions about residential segregation and considers population distributions by religion. The chapter concludes with some summary remarks on the overall socioeconomic and demography trajectory of Northern Ireland between 1971 and 2001.

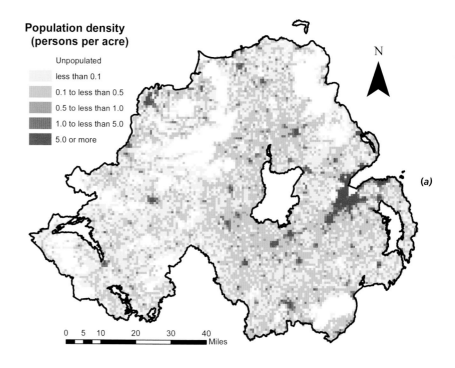

**Population density
(persons per acre)**

Unpopulated

less than 0.1

0.1 to less than 0.5

0.5 to less than 1.0

1.0 to less than 5.0

5.0 or more

N

(a)

0   5   10   20   30   40
Miles

Northern Ireland: A Unique Society?

Industrial decline and sectarian conflict resulted in the Northern Ireland
of 2001 being a very different place from the Northern Ireland of 1971.
Initially, we explore some of the many commonalities that Northern Ireland has with the rest of the world in terms of demographic change and
deindustrialization before pursuing the themes of segregation and religious
demography, which set Northern Ireland apart.

Figure 9.1 shows Northern Ireland's population density in 1971, 1991,
and 2001 as recorded for 1 km grid squares. The series does not include
1981 because this census took place against the backdrop of the unrest that
accompanied the Hunger Strikes and suffered from particular problems
of underenumeration, especially in Catholic areas.[3] The maps give a perspective to mapping different from that seen elsewhere in this book. The
areas that are empty, for example, are just that: they represent places such
as mountains where there is no population. The maps clearly show how important Belfast and its suburbs are as a center of population. The remainder
of Northern Ireland consists of Londonderry/Derry in the northwest and
a variety of smaller towns. The western part of Northern Ireland is more
sparsely populated than the east, although even in the east, areas such as
the Glens of Antrim in County Antrim and the Mourne Mountains in
County Down have large unpopulated areas.

Over the course of three decades the population of Northern Ireland
increased from 1.54 million in 1971 to 1.69 million in 2001, an increase of
around 10 percent.[4] Figure 9.2 shows that there was, however, a clear geographical pattern to this increase, with losses taking place in central parts
of Belfast and Londonderry/Derry and in the centers of some smaller towns
such as Coleraine and Ballymena, but with areas of population gain in close
proximity to these areas. This is an example of Northern Ireland exhibiting
the same types of social and geographical patterns as are found elsewhere.

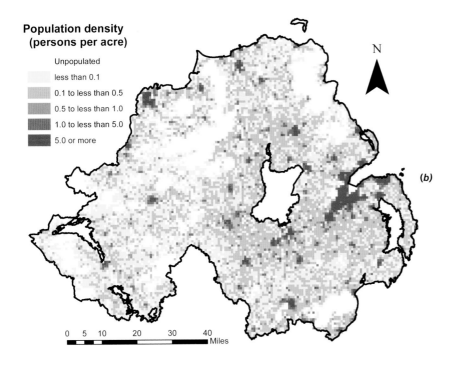

(b)

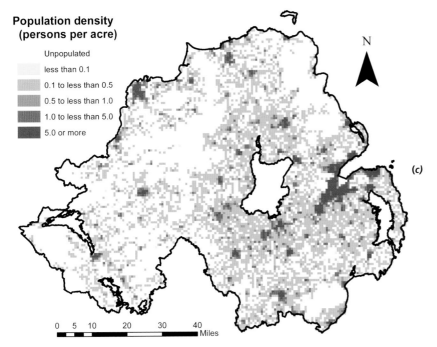

(c)

Typically, cities and towns throughout Europe and North America have lost population from their centers and gained it in their suburbs or other nearby rural areas, a process known as counterurbanization.[5] The impact of this on Belfast will be dealt with in chapter 11. The pattern of population loss from areas that were formerly densely populated, and population gain in less populated locales is summarized in figure 9.3, which emphasizes that those areas that were densely populated in 1971 tended to lose population over the next three decades. The other interesting point about this graph, however, is that, unlike in the Republic, there is little evidence of rural population loss; in fact, if anything, sparsely populated rural areas

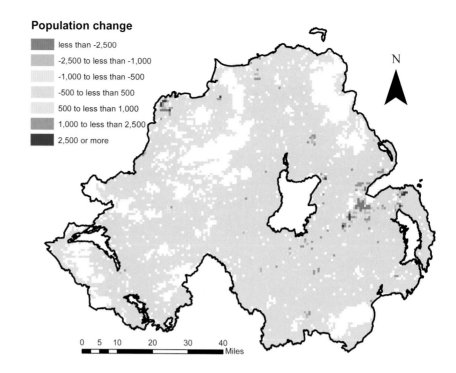

Fig. 9.2. Population change in Northern Ireland, 1971–2001.

Population change

- less than -2,500
- -2,500 to less than -1,000
- -1,000 to less than -500
- -500 to less than 500
- 500 to less than 1,000
- 1,000 to less than 2,500
- 2,500 or more

N

0 5 10   20   30   40
Miles

Fig. 9.3. Scatter plot comparing population in 1971 with population change between 1971 and 2001 for 1 km cells.

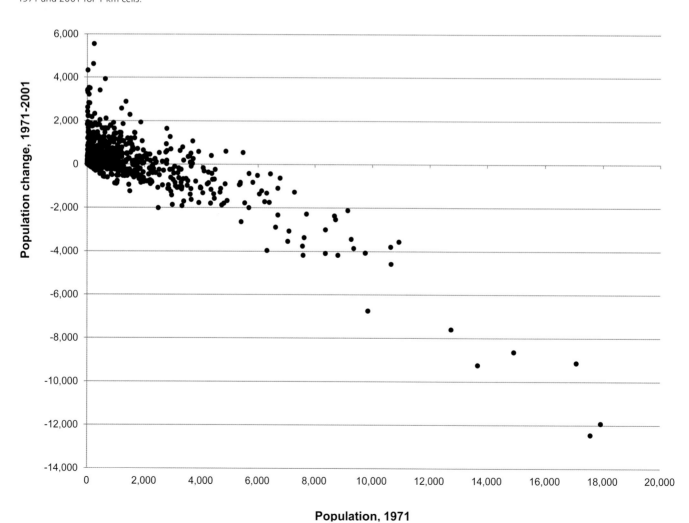

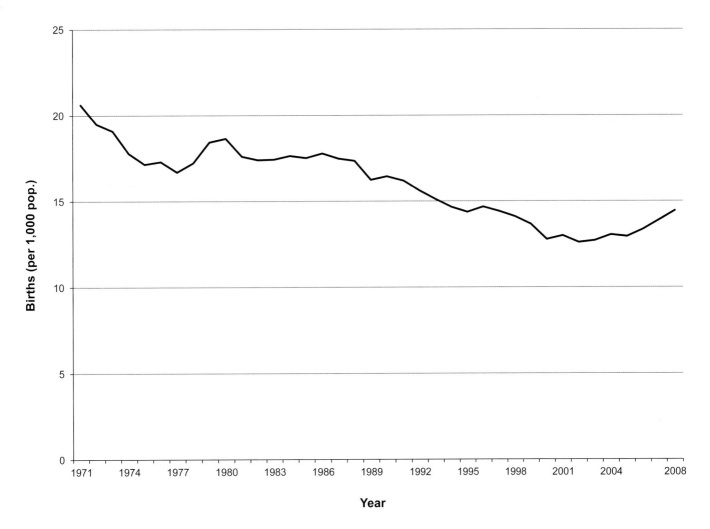

Fig. 9.4. The crude birthrate in Northern Ireland, 1971–2008. (Source: *Registrar General Northern Ireland Annual Report 2008* [Belfast: NISRA, 2009].)

tended to show gains, although figure 9.2 shows that, as with the Republic, the biggest gains tended to be near cities.

The similarities between Northern Ireland and other places in Europe go further than this. Historically, Northern Ireland has had a higher birthrate than many other parts of Europe, and this relatively high rate has persisted.[6] However, as figure 9.4 shows, there has been a long-term decline in the crude birthrate from 1971 until 2001 similar to that seen elsewhere. Indeed, it is only in the most recent period that this fall shows signs of reversing, and it is by no means certain that this move upward in birthrates will be long lasting. One consequence of this is that Northern Ireland has become a more aged society since 1971: over this period, the number of children fell by 17 percent, while the number of people of pensionable age rose by 39 percent.[7] Thus, although Northern Ireland's population grew between 1971 and 2001, this was mainly a result of declining mortality.

The aging society has been accompanied by a decline in the average size of household. In particular, the number of one-person households has grown: there are now more divorced and separated people, a greater number of widows and widowers, and, at the other end of the age spectrum, more single people who live alone. These trends are shown clearly in figure 9.5, which contrasts mean household size in 1971 and 2001. By 2001 there were many more areas with smaller mean household sizes than in 1971.

Fig. 9.5. Average household size in Northern Ireland in (*a*) 1971 and (*b*) 2001.

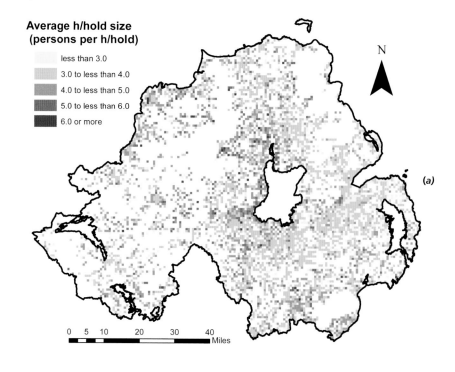

Average h/hold size
(persons per h/hold)

less than 3.0
3.0 to less than 4.0
4.0 to less than 5.0
5.0 to less than 6.0
6.0 or more

(a)

0 5 10   20   30   40
Miles

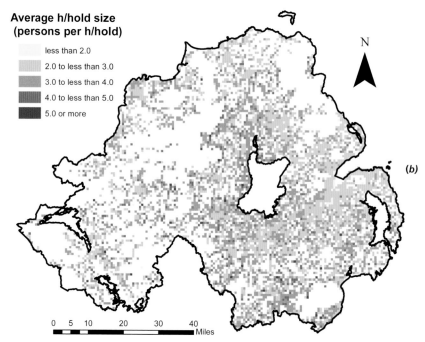

Average h/hold size
(persons per h/hold)

less than 2.0
2.0 to less than 3.0
3.0 to less than 4.0
4.0 to less than 5.0
5.0 or more

(b)

0 5 10   20   30   40
Miles

Economic and Social Change

The decline of heavy industry took Northern Ireland through a transition from the traditional manufacturing-based economy to what might be termed the "new economy."[8] This is reflected in the decline in both agriculture and manufacturing since 1971 and the growth of service-sector employment, particularly with the expansion of the public sector, the arrival of new service-sector employers such as call centers, and an increase in the financial sector. This has been matched by occupational changes.

Traditional occupations (e.g., manufacturing-related manual work) have declined at the expense of growth in "new jobs" in service employment. In some ways these trends have been similar to those in the Republic, but their impact has been different. The previous importance of manufacturing in Northern Ireland meant that its loss was traumatic, while the increase in service-sector jobs has been less dramatic and far more based around the public sector than south of the border. These changes have been accompanied by changes in the gender distribution of economic opportunity: by 2001 there were many more women in the workforce than in 1971. So the watchwords for the Northern Ireland economy are *manufacturing decline, service growth*, and *feminization*. This package of changes in the economy and labor market can be linked together as a narrative using concepts such as the transition from Fordism to post-Fordism.[9] At the heart of this concept and, indeed, others like it are notions of "before and after" and of wide-ranging fundamental changes in the way that the economy works.

While these changes have been traumatic for Northern Ireland's economy and society, and many people and places have lost as a result, others have gained. These patterns can be identified from the maps in figure 9.6, which show employed people as a percentage of the economically active population in 1971 and 2001. Two features stand out from these maps. The first is that employment rates were generally lower in Belfast and Londonderry/Derry in 2001 than they were in 1971. The second is the trend toward greater unevenness in the geography of employment in 2001 than in 1971. Despite this, one net outcome of the various forces acting upon Northern Ireland has been increasing affluence. Figure 9.7 shows that the proportion of houses that were owner occupied grew between 1971 and 2001, while rented accommodation of various kinds decreased. Some of this can be attributed to housing and planning policy, including initiatives such as "right to buy"—allowing tenants to purchase socially rented housing stock—but it is also highly likely that growing affluence has meant more owner occupation. Likewise, as shown in figure 9.8, private car ownership has become far more widespread.

To what extent, then, has Northern Ireland been a unique society in this period? The analysis presented so far indicates that in important respects it was far from unique. Trends such as counterurbanization, an ageing population, and increasing material affluence are typical of the developed world in this period, while the decline of traditional heavy industries and the difficulties in replacing them with service-sector jobs is typical of similar areas in the U.K. and beyond. In this regard the Republic was perhaps the exception, while Northern Ireland represents the rule. This account, however, ignores the themes of religious demography and residential segregation. It is therefore to these spatioreligious issues that we now turn to explore the question of whether Northern Ireland became a more segregated society between 1971 and 2001.

Religion: A More Segregated Society?

As chapter 6 established in its discussion of Partition, the "numbers game" in religious demography has always been important in Northern Ireland.

Fig. 9.6. Employment rates among the economically active population in (a) 1971 and (b) 2001.

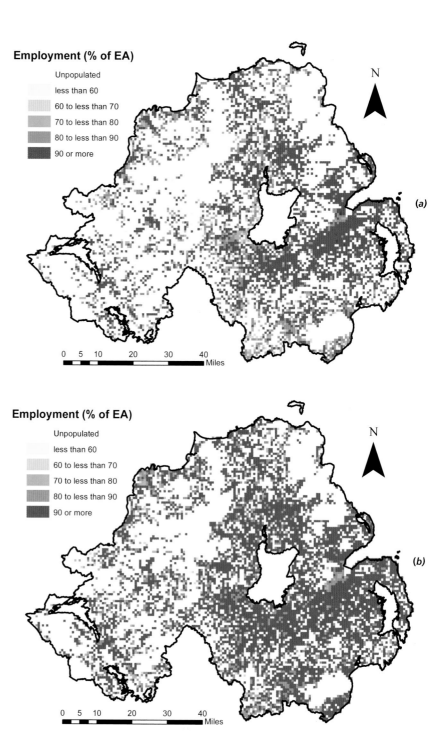

Employment (% of EA)

Unpopulated
less than 60
60 to less than 70
70 to less than 80
80 to less than 90
90 or more

0   5   10   20   30   40
Miles

(a)

Employment (% of EA)

Unpopulated
less than 60
60 to less than 70
70 to less than 80
80 to less than 90
90 or more

0   5   10   20   30   40
Miles

(b)

The state was set up to have a sustainable Protestant majority, with the six-county arrangement giving a balance of approximately one-third Catholics to two-thirds Protestants when it was formed. Since the 1970s there has been a widespread perception, seen often in the media, that Protestants are in demographic and economic retreat, while Catholics are advancing.[10] The prospect of a Catholic majority is viewed with hope by some and with fear by others.[11] Evidence from the census suggests that these perceptions may have some justification. In 1971 Catholics made up 34 percent of the population, virtually the same number as in 1911; however, by 2001 this had risen to either 47 or 45 percent depending on the variable used. The

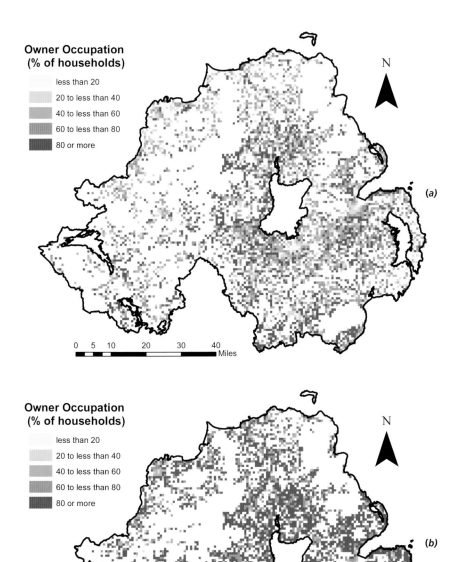

**Owner Occupation
(% of households)**

less than 20
20 to less than 40
40 to less than 60
60 to less than 80
80 or more

N

(a)

0 5 10 20 30 40
Miles

**Owner Occupation
(% of households)**

less than 20
20 to less than 40
40 to less than 60
60 to less than 80
80 or more

N

(b)

0 5 10 20 30 40
Miles

Fig. 9.7. Owner-occupied housing in (a) 1971 and (b) 2001.

reason for having two figures is that the 2001 census asked two questions on religion, one of which asked the person to identify his or her community background and the other to identify his or her religious affiliation. The percentages given above refer to community background and affiliation, respectively. Regardless of which is used, it is clear that the Catholic share of the population had grown significantly.

The fact that there were alternative questions in 2001 is perhaps indicative that, as in the Republic, the role of religion within Northern Irish society was changing. The religion question in the Northern Ireland census is voluntary, and there have been varying levels of nonresponse to it over time

Fig. 9.8. Households lacking a car in (a) 1971 and (b) 2001.

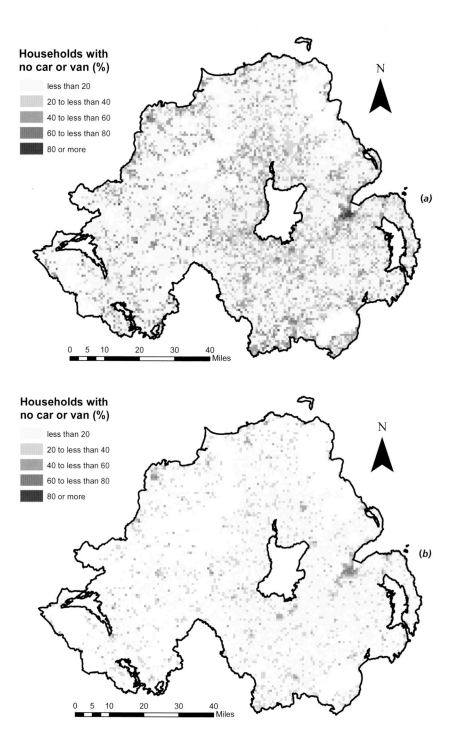

and between different groups. In 1971 the 7 percent nonresponse was primarily in Catholic areas. In 1991 the 9 percent nonresponse was primarily in Protestant areas, and in 1981 it reached 20 percent of the population, probably concentrated in Catholic areas.[12] The reasons for these variations are more complex than in other societies and may reflect fear or opposition to perceived government surveillance in addition to more conventional factors such as a decline in the importance of religion with secularization and modernization. Equally, the desire by some politicians on both sides to make their populations seem as large as possible may have encouraged people to identify themselves as religious when their "religion" was more concerned with community identity than faith. The 2001 community background question was, in part, a response to these problems. It asked about the religion that someone was brought up in rather than the religion he or she professed at the time of the census. It also imputed values for those who refused to answer the question. The community background question of 2001 followed the option of describing oneself as having no religion, which was first provided in 1991. The fact that both north and south added options that allowed respondents to have no religion suggests that secularization was rising across Ireland. Unlike the Republic, however, Northern Ireland did continue to divide Protestants into denominations such as Presbyterian, Church of Ireland, and Methodist. Thus in the Republic census questions on religion became less detailed, while in the north they became increasingly complex and separated religiously defined community identity from religious affiliation. This says much about the different ways in which religion was changing in the two parts of the island and the political context in which this was occurring.

The changing geography of religion is shown in figure 9.9, which maps the numbers of Catholics in 1971, 1991, and 2001. It shows a clear pattern of concentration in parts of Belfast, south Down, and south Armagh and in areas west of the River Bann, especially in and around Londonderry/Derry. There were also large areas with very low Catholic populations, especially in north Down and Antrim. Figure 9.10 shows the same pattern for Protestants, who were very much concentrated in north Armagh, north Down, and east Antrim.

Figure 9.11 explores spatioreligious change over this time. It is clear that parts of central Belfast and the east bank of the Foyle in Londonderry/Derry lost large numbers of Catholics over this period. The majority of grid squares, however, have seen moderate increases in their Catholic populations, but in some places—such as south and west Belfast and Londonderry/Derry

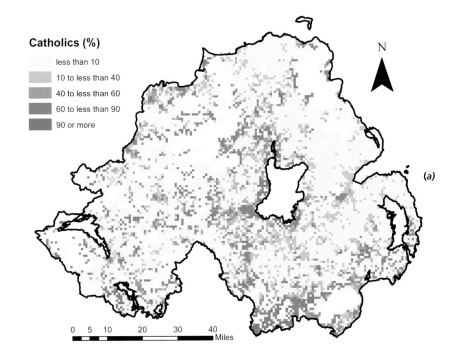

Fig. 9.9. The distribution of Catholics in (a) 1971, (b) 1991, and (c) 2001. Earlier maps from this series are available in figure 7.11. (*Continued on the next page*)

Fig. 9.9. The distribution of Catholics in (*b*) 1991, and (*c*) 2001. Earlier maps from this series are available in figure 7.11.

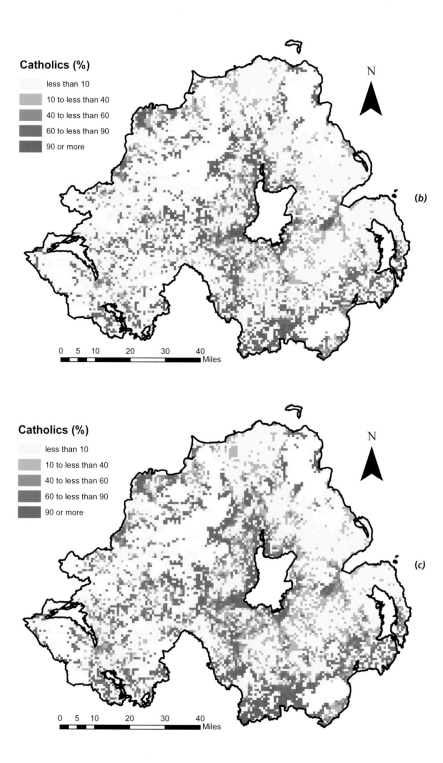

Catholics (%)

less than 10
10 to less than 40
40 to less than 60
60 to less than 90
90 or more

(b)

0 5 10   20   30   40
Miles

Catholics (%)

less than 10
10 to less than 40
40 to less than 60
60 to less than 90
90 or more

(c)

0 5 10   20   30   40
Miles

west of the Foyle—these gains have been large. The pattern for Protestants shows some interesting contrasts to this. There are different geographies of population loss, with a far more general decline in Belfast. Fewer grid squares saw Protestant population growth, and, rather than these being in the south and west of Belfast, the grid squares near Belfast that show Protestant growth tended to be more toward the east. Around Londonderry/Derry there was Protestant decline to the west of the Foyle and growth to the east, the opposite of the picture for Catholics. These themes of differential population increase and decrease by religion in Belfast will be taken up in more depth in chapter 11. For the moment, however, it is sufficient to

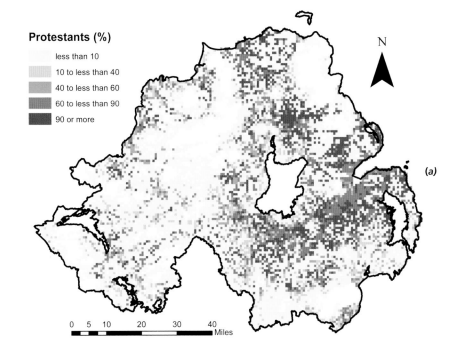

Protestants (%)

less than 10
10 to less than 40
40 to less than 60
60 to less than 90
90 or more

0 5 10 20 30 40 Miles

(a)

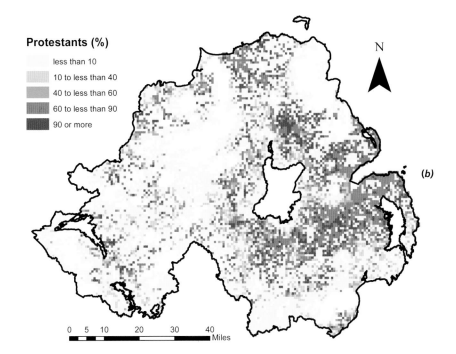

Protestants (%)

less than 10
10 to less than 40
40 to less than 60
60 to less than 90
90 or more

0 5 10 20 30 40 Miles

(b)

Fig. 9.10. The distribution of Protestants in (*a*) 1971, (*b*) 1991, and (*c*) 2001. Earlier maps in this series, showing Presbyterians and the Church of Ireland, are shown in figures 7.12 and 7.13, respectively. (*Continued on the next page*)

note that the evidence on population change is mixed. In some respects, most notably the declines for both groups in Belfast, the geography of population change is similar, but in terms of the geography of where the two communities were increasing there are important differences.

The second major discourse around population, particularly near the time of censuses, has been one of "increasing apartheid." This was important in the years around 1991 and 2001, and this story has been repeated on many occasions more recently in the media with arguments that already high segregation is increasing. Debates about segregation have been tied into criticisms of the Good Friday or Belfast Agreement of 1998, the major

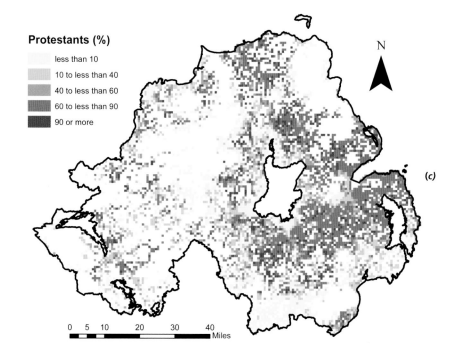

milestone in the peace process. It has been argued that worsening of community relations—evidenced by increasing segregation as the communities "choose to live apart"—has been caused by flawed political institutions.[13]

I. Shuttleworth and C. Lloyd calculated segregation measures using the 1 km grid square data, and these are summarized in table 9.1.[14] Their analysis suggests that the discourse of increasing apartheid exaggerates the amount of recent growth in residential segregation. Far from seeing "inexorable growth," the census shows that the majority of the growth in residential segregation in Northern Ireland occurred between 1971 and 1991. The dissimilarity index rose from 56 percent in 1971 to 66.8 percent in 1991 and then to either 67 or 64.9 percent in 2001 for religion and community background, respectively. The isolation indexes largely follow this trend. In both cases the bulk of the segregation increase occurred in the 1971 to 1991 period, but, in contrast, between 1991 and 2001 levels of segregation remained the same or, depending on the index, decreased slightly. In some ways, therefore, the commentary after 2001 that talked up increasing segregation was reporting "old news" from the 1970s and 1980s that had not been recognized until the Northern Ireland grid square data had been analyzed.

Although the census shows geographical patterns and can be used to calculate segregation indexes, it lacks data to help in the interpretation of these patterns, and it says nothing directly about the causality of population change. This has to be inferred and is reliant upon analytical judgment. Given these caveats, what can be said about the increase in segregation between 1971 and 2001 and the geography of population change by religion in Northern Ireland? First, it is probable that forced housing moves, particularly at the start of the period, contributed both to differential geographies of population increase and decrease and to increases in segregation.[15] Second, it is unlikely that this tells the full story, and hence there is a need to look at other factors that shape population. Likely candidates include housing policy, employment change, and counterurbanization—common

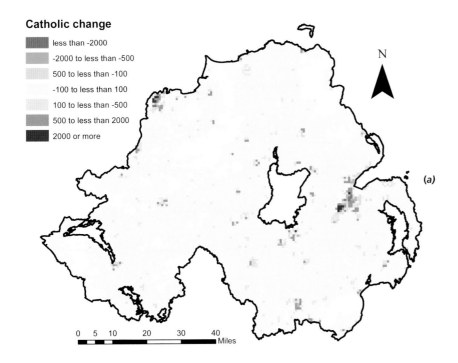

Catholic change

- less than -2000
- -2000 to less than -500
- 500 to less than -100
- -100 to less than 100
- 100 to less than -500
- 500 to less than 2000
- 2000 or more

N

(a)

0 5 10   20   30   40
                          Miles

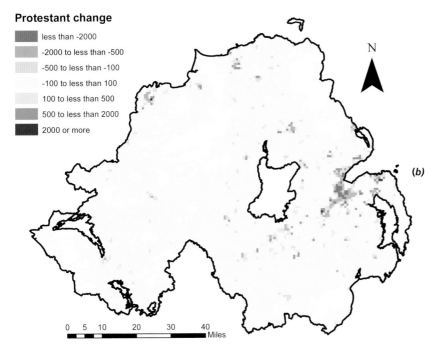

Protestant change

- less than -2000
- -2000 to less than -500
- -500 to less than -100
- -100 to less than 100
- 100 to less than 500
- 500 to less than 2000
- 2000 or more

N

(b)

0 5 10   20   30   40
                          Miles

Table 9.1. Indexes of segregation for Northern Ireland, 1971–2001

	1971	1991	2001 (R)	2001 (CB)
Index of dissimilarity	56.0	66.8	67.0	64.9
P* Catholics	61.2	73.8	75.7	73.2
Expected P* Catholics	34.2	43.0	46.8	44.8
Difference	27.0	30.8	28.9	28.4
P* Protestants	79.8	80.2	78.6	78.0
Expected P* Protestants	65.9	57.0	53.2	53.1
Difference	13.9	23.2	25.4	24.9

Note: Figures for 2001 refer to religious affiliation (R) and community background (CB).

factors in Europe and North America—together with the continued growth of the Catholic population. Shuttleworth and Lloyd suggest that the major driver of increasing segregation has been growth in the geographical concentration of the Catholic population.[16] Following this argument, figure 9.11 begins to show where this concentration has happened and enables some speculation about how it happened. One feature is that the west of Northern Ireland has become increasingly Catholic through time as the overall number of Catholics has grown. A second feature has been the differential geography of population increase and decrease in and near Belfast and Londonderry/Derry. In particular, the Catholic population growth to the south and west of Belfast is noteworthy and is associated with the large public housing estates of the 1970s and 1980s. These areas, which were formerly thinly populated with Protestants or started with mixed populations, became centers for Catholic population growth.[17] In contrast, Protestants became more widely geographically dispersed because they had greater opportunity to move outside Belfast within the owner-occupied housing market. Exploratory analyses suggest that Catholics and Protestants moved away from the same types of areas.[18] Typically, these were inner-city, socially deprived, densely populated places, but there were unequal responses to the same forces between Catholics and Protestants because of the inequalities between the two communities at the start of the period.

Conclusions

Northern Ireland changed in complex ways in the thirty years between 1971 and 2001. During these years the image of Northern Ireland was dominated by communal conflict and violence; however, as this chapter has shown, it experienced many of the changes seen in other societies. One important conclusion, then, is that Northern Ireland is not a place apart; indeed, from an economic perspective, the Republic was exceptional, while Northern Ireland's experience was typical of many other parts of the developed world. Beyond this, the conflict, and the understandable emphasis on communal division, does set Northern Ireland apart. There is clear evidence for Catholic demographic advance and for increased spatioreligious segregation through time. Claims that segregation grew inexorably between 1971 and 2001 exaggerate the reality, which saw residential segregation peak in 1991 and remain stable or decrease slightly by 2001. Indeed, it seems likely that much of the growth in segregation took place during the early 1970s in the early stages of the Troubles and some unmeasured part in the late 1960s before the 1971 census. Exact figures for this precensus period are hard to come by, but riots in 1969 alone are believed to have displaced almost two thousand people, of whom around 80 percent were Catholic.[19]

The similarities and contrasts in the place of religion in society north and south of the border are also interesting. In the Republic religion has clearly become less important for reasons discussed in chapter 8, and the traditional split between Catholic and Protestant has become less important for a number of reasons, including the relatively small Protestant minority, an influx of immigrants who do not fit the traditional Irish Catholic and Protestant divisions, and a decline in the importance of the link between religion, identity, and politics. These reasons may have been further exacerbated by the economic success of the country at this time. In the north, religion also seems to have become less important, as people increasingly do not identify with one religion or the other. Nevertheless, the impact of religion, in the widest sense of the term, has of course been huge. Religious division, community identity, and the politics of republicanism/nationalism, on the one hand, and loyalism/unionism, on the other, have come together to cause widespread violence. The economic stagnation of Northern Ireland during the 1970s and 1980s will have made matters worse, as unemployment and alienation from mainstream society of some groups led to further tensions. These economic difficulties were further exacerbated by widespread discrimination against Catholics, especially in the early part of this period.[20] As always in Northern Ireland, the reality is more complex than the way in which it has been portrayed in many accounts. This complexity will be a feature of the next two chapters on the recent past in Northern Ireland.

10 Communal Conflict and Death in Northern Ireland, 1969 to 2001

The conflict in Northern Ireland known as the Troubles started in the late 1960s and largely ended following the Good Friday or Belfast Agreement of 1998, although a decade later violence continued to occur, albeit at a much reduced level. The violence led to over 3,500 deaths. This could be argued to be a small figure, far outweighed in importance by other causes of death such as cancer and heart disease. Even as a percentage of the population it may seem small, coming to only 0.23 percent of Northern Ireland's population. If this seems like a small figure, however, its numerical significance can be shown by calculating what this would mean if it were applied to British or U.S. populations. A similar death rate in Britain would lead to approximately 130,000 deaths, which equates to the loss of a town like Brighton or Peterborough. In the United States, with its larger population, the equivalent would be approximately 500,000 deaths, comparable to total U.S. military deaths in World War II. From this perspective, it is clear that conflict and violence have led to significant numbers of deaths in Northern Ireland. Moreover, as we will demonstrate, conflict-related killings were geographically concentrated in certain places, including parts of Belfast, some sections of mid-Ulster, and rural areas near the border such as south Armagh. This meant that the direct traumatic impact of the conflict was disproportionately felt by a relatively few communities.

This unevenness of experience has heightened and concentrated the local consequences of violence. Some communities are still struggling to overcome the legacy of historic conflict and ongoing sectarianism, itself a current cause of bad community relations and an overhang from the past. Other communities have only felt the results of violence indirectly. Thus, in addition to the geographies of religion, identity, and politics identified in much of this book, this chapter explores a directly linked geography of violent death.

Not surprisingly, given the numerical weight of violent deaths and their ongoing legacy for families and communities, interpretations of the conflict are controversial. Understanding the conflict is made more difficult because it was not a simple and two-sided dispute between republicans (Catholics trying to create a united Ireland independent from Britain) and loyalists (Protestants in favor of preserving Northern Ireland as part of the U.K.). The British state was also a participant in the conflict and was not a neutral third party, although it often would have liked to have been seen in this light. Even the terms used to describe the major protagonists are sometimes contested. "Loyalist," for example, tends to be used to describe Protestant paramilitaries but would not usually be used for unionist Protestants as a whole, reflecting the uneven social and spatial incidence of violence.

Similarly, "republican" tends to only refer to groups who were prepared to use violence to further their aim of a united Ireland rather than the more moderate opinions held by much of the nationalist Catholic population.

In many respects the Troubles involved two major separate but related conflicts and several more minor ones.[1] The first major conflict involved fighting between republicans and the security forces.[2] The second conflict comprised loyalists primarily using violence against Catholic civilians.[3] At the same time, paramilitaries on both sides used violence to defend their own communities from other paramilitaries. Additionally, feuding between different factions of both republicans and loyalists meant that not all violence was external—directed at members of the "other" group—but was sometimes internal—loyalists attacking other loyalists or republicans attacking other republicans. Civilians were often casualties in most, if not all, of these conflicts. Sometimes this was the result of them being bystanders, but at other times there was more direct targeting. Riots and other street disturbances often resulted in the security forces killing civilians, particularly Catholics. Paramilitaries on both sides killed civilians on the other side either as a result of deliberate targeting or simply because they were bystanders. Additionally, paramilitaries killed civilians on their own side, sometimes accidentally but also deliberately in acts of "internal policing." For much of the conflict both loyalist and republican paramilitaries acted as the de facto police forces within their own communities, imposing discipline in areas where they believed that the state had no legitimacy or authority. These paramilitaries delivered their own form of justice in the manner of "punishment beatings," which sometimes proved fatal. This internal policing was not aimed only at informers; petty criminals and people deemed to be antisocial were also frequent victims.[4]

A Database of Deaths during the Troubles

This chapter is based on the georeferenced database of killings compiled by Malcolm Sutton and hosted by the University of Ulster's Conflict Archive on the Internet (CAIN).[5] This database was initially created to act as a memorial to those who had died during the conflict and formed the basis for a book, *Bear in Mind These Dead: An Index of Deaths from the Conflict in Ireland 1969–1993*.[6] The process of updating the database has continued to the present to include new deaths as they occur.[7] While the rate of entries has declined substantially since the Good Friday Agreement, the process of keeping the information up-to-date and, more specifically, of correcting the information on individuals already within the record continues.

Despite the decline in the rate at which new names are added, the fact that the Sutton database continues to grow reflects part of the ongoing, unfinished business of violent conflict in Northern Ireland and means that we had to set a time limit on entries in order to perform our analysis. This chapter analyzes the period that runs from the urban disturbances arising out of the civil rights struggle of the summer of 1969, which marked the onset of the Troubles, through to the end of 2001. During this period the database records 3,524 deaths, of which 3,268, or 93 percent, occurred within Northern Ireland. Of the remaining 256 deaths, 125 occurred in Britain, 113 in the Irish Republic, and 18 in mainland Europe. While deaths outside of Northern Ireland often achieved far higher levels of international attention, one of the core objectives of IRA attacks abroad, violence was actually an overwhelmingly Northern Ireland phenomenon.

The Sutton database provides information on the name of each person killed, the date of death, and the person's age and gender. It also contains information on the background of victims, including their religion and whether they were members of the security forces or a paramilitary group. For most of the chapter, the broad classes of *security forces* (a category that mainly consists of the British Army, the Ulster Defence Regiment, and the Royal Ulster Constabulary), *republican paramilitary, loyalist paramilitary,* and *civilian* will be used, with civilians being further subdivided into *Catholics, Protestants,* and, where appropriate, *not from Northern Ireland* or *other*. Information is also given on the perpetrator of the attack, which we have typically classified as *security forces, republican paramilitary,* and *loyalist paramilitary. Irish security,* which refers to the Irish police force, the Garda Síochána, and the Irish Army, is also sometimes included. The database also provides a textual description of where each killing took place, which was used to allocate each killing to a location. Locations are more precise in large urban areas such as Belfast than in some rural areas. There are also questions about the appropriate spatial scale for the analysis—in Belfast, for instance, spatial differences between one side of the street or the other can be highly significant in some of the so-called interface areas where concentrations of the two communities are immediately adjacent to each other. In other places such precision would be spurious and unnecessary.

The Northern Ireland grid square product, described in chapter 9, provides contextual data. These data allow us to relate Troubles-related deaths to a population base and then compare deaths to the background demography. The present analysis is based on killings for the entire period of the Troubles; thus, we computed the average population for the four census years and used it to provide an average background population.

An Overview of Conflict-Related Deaths

The multifaceted nature of the conflict is apparent in table 10.1, which shows the perpetrators of violence and their victims as recorded by the Sutton database. Republican paramilitaries, in practice mainly the Provisional Irish Republican Army (usually simply known as the IRA), were the leading killers, being responsible for over two thousand deaths, 56 percent of the total. As might be expected given the political objectives of the republican campaign, just over 50 percent of the IRA's victims were from the security forces. Republicans were also responsible for 738 civilian deaths, which, at 36 percent of the total number of people killed by republicans, represent the second largest group of their victims. Fifty-four percent of these civilians were Protestants, many of whom are likely to have been claimed by republicans to be "legitimate targets" who were involved in some way in supporting British institutions. Others would simply have been bystanders. It is noteworthy that 228 Catholic civilians were also killed by republicans. Some of these victims would also have fallen into the category of "legitimate targets," while others would have been a result of "internal policing."

Table 10.1. Number of persons killed by organization responsible

		Perpetrator					
		Security forces	Loyalist paramilitary	Republican paramilitary	Irish security	Not known	Total
Victim	Security forces	13	14	1,078	0	7	1,112
	Civilian:	190	873	738	0	56	1,857
	Catholics	162	686	228	0	26	1,102
	Protestants	24	132	402	0	26	584
	Not from Northern Ireland	4	55	108	0	4	171
	Loyalist paramilitary	14	91	45	0	1	151
	Republican paramilitary	145	42	185	5	17	394
	Irish security	0	10	0	0	0	10
	Total	362	1,020	2,056	5	81	3,524

Additionally, 185 republican paramilitaries were recorded as being killed by republicans, almost the same number of republicans as were killed by the security forces and loyalist paramilitaries combined. This suggests that accidental deaths—particularly resulting from the use of homemade bombs, intergroup feuding, "internal policing," and "disciplinary actions" were a greater menace to life and limb for republicans than their enemies were.

The next largest number of deaths was caused by loyalist paramilitaries. Here, the pattern of victims shown in table 10.1 indicates a very different kind of conflict. Since they were not politically opposed to the security forces (in fact, there have been claims that there was collusion between the two), it is noticeable that loyalists were responsible for only a very small number of security force deaths. Eighty-five percent of their victims were civilians, 79 percent of whom were Catholic. Thus it is clear that much of this violence was deliberately aimed at Catholic civilians, with loyalists being generally acknowledged to have a far wider definition of "legitimate target" than republicans. Loyalists did, however, share one characteristic with their republican counterparts: they were far more dangerous to their own side than to opposing paramilitaries to the extent that 60 percent of loyalist victims were killed by other loyalists. Many of these deaths can be attributed to feuding between and within the various loyalist groups.

The security forces are in third place in the number of deaths they were responsible for. Fifty-two percent of their victims were civilians. The asymmetric role of the security forces is revealed by the religious balance of these civilians, with the number of Catholic civilians killed by the security forces outnumbering those of Protestants by almost seven to one. A similar trend is also found when deaths of republican and loyalist paramilitaries caused by the security forces are considered: republican deaths outnumber loyalists by over ten to one.

Figure 10.1 shows how the numbers of killings varied over time. The pattern is uneven: after the start of the Troubles in 1969 violence grew rapidly. The number of deaths peaked in 1972 with 479 deaths. Fatalities then dropped sharply to around 250 per year, and in the mid-1970s they dropped again to remain at between 50 and 100 per year for much of the late 1970s,

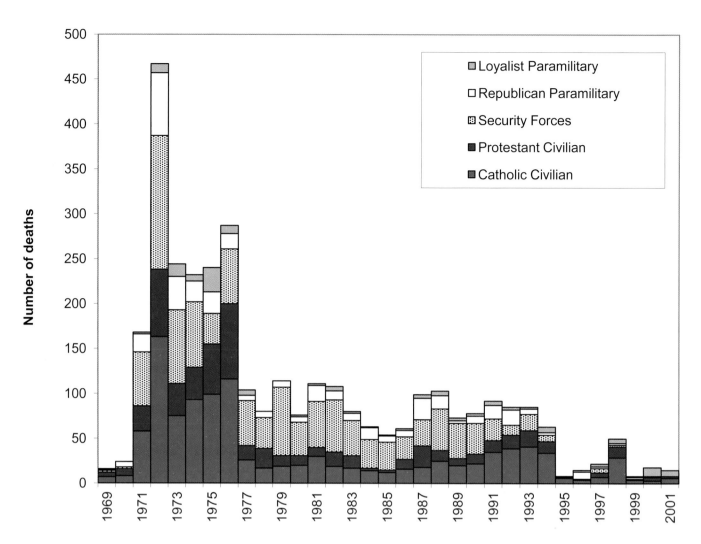

Fig. 10.1. Numbers of deaths per year during the Troubles.

1980s, and early 1990s. Another sharp drop occurred after the paramilitary ceasefires of 1994 and 1996. Interestingly, the proportions of deaths between the different groups remain broadly constant through much of the period.

Figure 10.2 shows the locations of deaths per 1 km grid square. The main feature here is the spatial concentration of conflict-related deaths in Belfast, reflecting both population numbers and the central importance of Belfast as an arena for violence. Figure 10.3 standardizes these data using the underlying population to give deaths per capita and gives a different picture. Belfast is still important but less so than before; however, the border area of south Armagh becomes clearer as a focus of violence, as do Londonderry/Derry and rural areas such as east Tyrone and north Armagh.

Complex Geographies of Conflict-Related Death

The statistics above are useful as a general introduction to the spatial and temporal incidence of violent deaths in Northern Ireland. To understand more fully the multifaceted nature of the conflict, it is necessary to drill down into the data and to disaggregate them by perpetrator and victim to explore the often complex geographies of Troubles-related deaths. These geographies are an expression of the various political strategies of participants in the conflict and the ways in which territoriality was used to control,

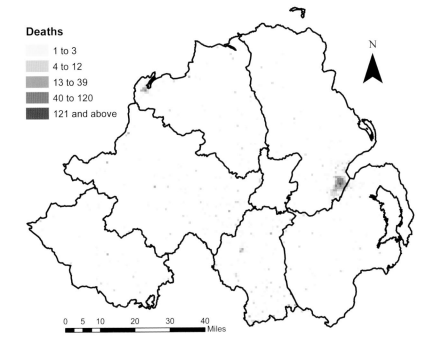

Deaths

1 to 3
4 to 12
13 to 39
40 to 120
121 and above

0 5 10   20   30   40
Miles

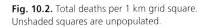

Fig. 10.2. Total deaths per 1 km grid square. Unshaded squares are unpopulated.

mark, and delimit populations by republican and loyalist paramilitaries and the British state.[8]

Deaths in the Troubles in Northern Ireland did not happen at random. They were, to a greater or lesser extent, the products of the strategies of the paramilitaries and the security forces to extend or protect territorial control and to work toward political goals. Territory is important in understanding political strategies and in interpreting the geographies of violent death. Northern Ireland is a communally divided society where competing British and Irish nationalisms throw the state's continued existence and

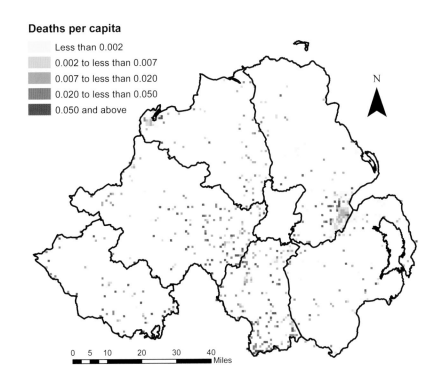

Deaths per capita

Less than 0.002
0.002 to less than 0.007
0.007 to less than 0.020
0.020 to less than 0.050
0.050 and above

0 5 10   20   30   40
Miles

Fig. 10.3. Deaths per head of population during the Troubles. This is calculated as total killings in each 1 km square (figure 10.2) divided by mean population from the censuses of 1971 to 2001. Only cells containing at least one killing are shown.

Fig. 10.4. The distribution of all killings by republican paramilitaries. The pattern has been smoothed using density smoothing; the legend uses a geometric progression.

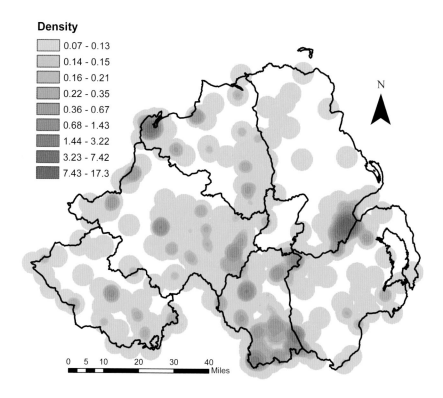

Density

	0.07 - 0.13
	0.14 - 0.15
	0.16 - 0.21
	0.22 - 0.35
	0.36 - 0.67
	0.68 - 1.43
	1.44 - 3.22
	3.23 - 7.42
	7.43 - 17.3

N

0 5 10   20   30   40
Miles

constitutional status into doubt. While these disputes have a territorial dimension at the level of Northern Ireland as a whole, where the question is about its constitutional status as part of the U.K. or as a united Ireland, the disputes also have implications at lower spatial scales where territory becomes a proxy for political and cultural control and where it can also be used as a marker for ethnonational identity.[9] Territory can often be seen as a zero-sum game in which a bigger share for one group means less for the other and "advance" for some means "retreat" for others. Local territories can be viewed as the "national question" in microcosm; thus, local territories and spatioreligious patterns of segregation are imbued with a political significance that is hardly surprising, given the politicization of Northern Ireland's religious demography. Territoriality can be viewed in a variety of often contradictory ways. High levels of local segregation can be interpreted as being harmful, hindering social integration and contributing to continued intergroup conflict. Equally, homogeneous territorial blocks can also be seen as being places of safety where residents can feel secure from attack. Territory can also sometimes be seen as a "prison" where residents are not safe, least of all from their "own side."[10]

Figure 10.4 shows the distribution of killings by republicans during the Troubles. The map is based around the locations from the Sutton database, but a technique called *density smoothing* has been used to simplify the pattern and make it more understandable cartographically. This technique involves counting the number of events, in this case, killings, that occurred near to each location, with nearer events having more impact on the density than those that are farther away. This technique has the effect of blurring the pattern slightly, but it is useful in identifying clusters where many events have occurred near to each other and in distinguishing places where multiple events have occurred at a single location from those where only a single event occurred.[11] The map shows a distinct geography. The largest number of republican killings took place in Belfast and Londonderry/Derry. A more dispersed but highly significant cluster is found in the south Armagh and Newry area. Of the more localized clusters, some are largely based on a single incident. These include the 1998 Real IRA bombing of Omagh in the west of Tyrone, which killed twenty-nine people; the 1982 Droppin Well pub bombing in Ballykelly, northeast of Londonderry/Derry, in 1982 in which the Irish National Liberation Army (INLA) killed seventeen; the 1978 bombing of the La Mon restaurant east of Belfast by the IRA, which killed twelve; and the 1972 Claudy bombings southeast of Londonderry/Derry, which killed nine.[12] It is noticeable that these mass casualty events in areas remote from the majority of republican activity tended to kill civilians, although the Droppin Well bombing is the possible exception to this in that eleven of its seventeen victims were soldiers. Other clusters are based on larger numbers of incidents. These tend to be concentrated in the towns of north Armagh and east Tyrone, including Armagh City, Lurgan, Portadown, and Dungannon. They are also found in Strabane and Castlederg on the western border of Tyrone. The town of Enniskillen in Fermanagh also stands out as a cluster. Here twelve of the twenty-two deaths occurred in the 1987 Remembrance Day bombing by the IRA, while the remainder took place in more scattered incidents.[13]

Figure 10.5 subdivides deaths of the security forces and loyalists at the hands of republicans. Figure 10.5a shows the deaths of the security forces. Perhaps not surprisingly, as they account for around half of the republicans' victims, these deaths follow a pattern broadly similar to that of figure 10.4. It is noticeable, however, that border areas generally seem to have been dangerous to the security forces. Rural parts of east Tyrone, Fermanagh, and even east County Londonderry also appear relatively dangerous, in part because of the use of remote areas to conduct ambushes. Antrim and north Down were probably the safest places for the security forces, and it is no coincidence that these areas have the smallest Catholic populations. Figure 10.5b shows the highly contrasting pattern of republican killings of loyalist paramilitaries. There were only forty-five of these killings, compared to over a thousand killings of the security forces, and they show a very different geography being focused in three clusters: the Belfast area stretching southwest to Lurgan and the Maze Prison, Londonderry/Derry, and Strabane. In some ways this is misleading, as the Strabane "cluster" consists of only one death and the Londonderry/Derry cluster of two deaths. These only appear to be significant due to the small numbers overall.

Figure 10.6 shows the distribution of civilians killed by the republicans, subdividing them between Protestants (figure 10.6a) and Catholics (figure 10.6b). The Catholic pattern is concentrated in areas that had the highest Catholic populations and the highest levels of violence by republican paramilitaries. These include Belfast, south Armagh, east Tyrone, and Londonderry/Derry. The Protestant pattern is more dispersed and, as previously identified, has some major local clusters caused by single events, usually bombings in which large numbers of civilians were killed.

Figure 10.7 shows the locations of the 1,020 killings attributed to loyalist paramilitaries. The similarities and differences between this figure and

Fig. 10.5. The distribution of the killings of (a) the security forces and (b) loyalist paramilitaries by republican paramilitaries.

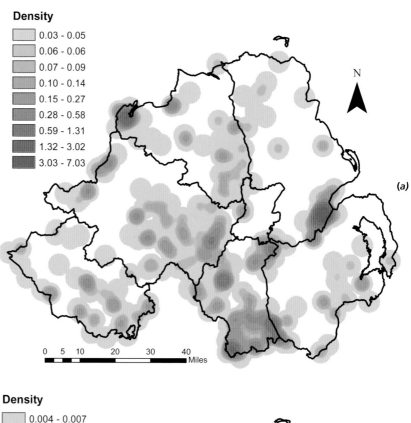

Density

	0.03 - 0.05
	0.06 - 0.06
	0.07 - 0.09
	0.10 - 0.14
	0.15 - 0.27
	0.28 - 0.58
	0.59 - 1.31
	1.32 - 3.02
	3.03 - 7.03

N

(a)

0 5 10   20   30   40
Miles

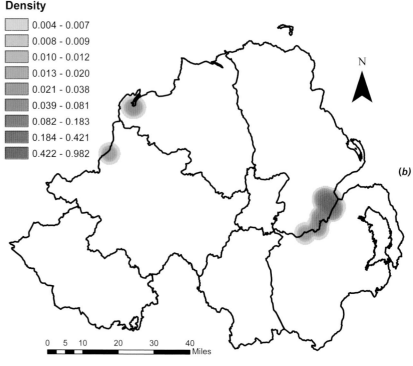

Density

	0.004 - 0.007
	0.008 - 0.009
	0.010 - 0.012
	0.013 - 0.020
	0.021 - 0.038
	0.039 - 0.081
	0.082 - 0.183
	0.184 - 0.421
	0.422 - 0.982

N

(b)

0 5 10   20   30   40
Miles

figure 10.4, which shows the killings by republicans, are interesting. As with republicans, many loyalist killings took place in and around Belfast, although, even from these maps, it is clear that republican action was focused on the west, while loyalists were more active in the north and east of the city. Londonderry/Derry is also a cluster but perhaps not as pronounced as for republicans. North Armagh and east Tyrone again come across as places with many killings, showing that these areas were contested by both sets of paramilitary groups. By contrast, many of the other places with clusters

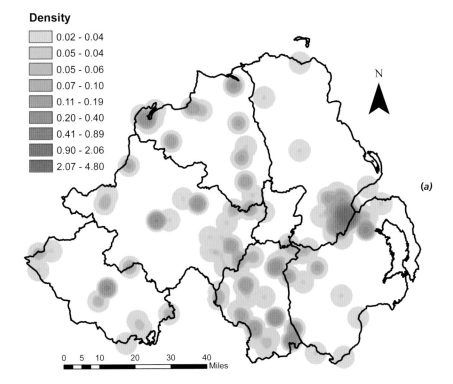

Density

- 0.02 - 0.04
- 0.05 - 0.04
- 0.05 - 0.06
- 0.07 - 0.10
- 0.11 - 0.19
- 0.20 - 0.40
- 0.41 - 0.89
- 0.90 - 2.06
- 2.07 - 4.80

N

(a)

0 5 10   20   30   40
Miles

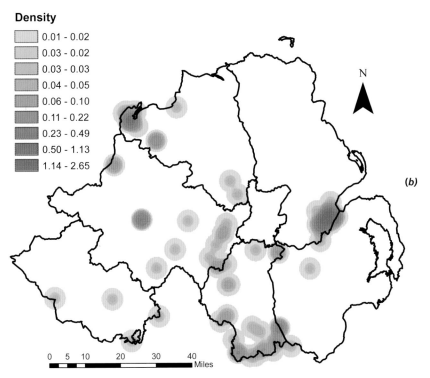

Density

- 0.01 - 0.02
- 0.03 - 0.02
- 0.03 - 0.03
- 0.04 - 0.05
- 0.06 - 0.10
- 0.11 - 0.22
- 0.23 - 0.49
- 0.50 - 1.13
- 1.14 - 2.65

N

(b)

0 5 10   20   30   40
Miles

Fig. 10.6. The distribution of killings of civilians by republican paramilitaries: (*a*) Protestants and (*b*) Catholics.

of killings by republicans have very few killings by loyalists. These include south Armagh and Newry, Fermanagh, and much of Tyrone. Large areas of County Londonderry, north Antrim, and south Down had few loyalist killings, which largely corresponds to the pattern for republicans. Another similarity with Republicans is the presence of isolated clusters that represent multiple killings in single events. These include the 1993 killings in the Rising Sun Bar in Greysteel northeast of Londonderry/Derry by the Ulster Freedom Fighters (UFF), which killed eight; the 1994 attack on

Fig. 10.7. The distribution of all killings by loyalist paramilitaries.

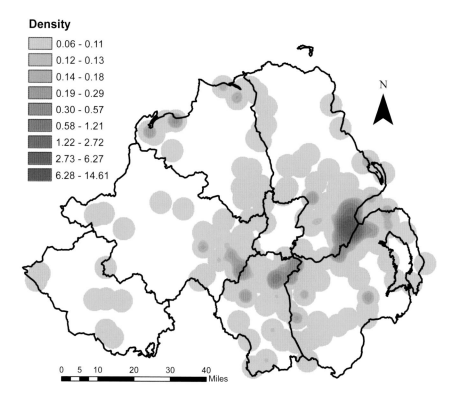

Density

- 0.06 - 0.11
- 0.12 - 0.13
- 0.14 - 0.18
- 0.19 - 0.29
- 0.30 - 0.57
- 0.58 - 1.21
- 1.22 - 2.72
- 2.73 - 6.27
- 6.28 - 14.61

N

0 5 10   20   30   40
Miles

the Heights Bar in Loughinisland in south Down, which killed six; and the shootings at Boyle's Bar in Cappagh in east Tyrone, which killed four. Both of these latter attacks were perpetrated by the Ulster Volunteer Force (UVF).[14] The major difference between these attacks and similar mass killings by republicans is that these were shootings rather than bombings. A similarity is that again mainly civilians were killed, except for the Boyle's Bar shooting, in which three of the four victims were members of the IRA.[15] These attacks and similar ones in more populated areas were in fact the bloodiest results of a loyalist strategy of deliberately targeting civilians in areas in which Catholics were known to congregate.

Figure 10.8 shows the distribution of loyalist killings of republicans. Of the forty-two of these, twenty-five occurred in and around Belfast, but none occurred in Londonderry/Derry. It is important to note that many of the remaining "clusters" are again only single deaths, and these are concentrated in north Armagh and east Tyrone. Most of the remainder occurred on or near the border between County Londonderry and Antrim. No map of loyalist killings of the security forces has been included, as there were only fourteen of these, of which ten occurred in Belfast.

Figures 10.9a and 10.9b show the killings of civilians by loyalists. Catholic civilians represent 67 percent of loyalist victims, so figure 10.9a is very similar to the overall pattern of loyalist killings shown in figure 10.7. The distribution of the 132 Protestant civilian deaths perpetrated by loyalists, 13 percent of their total, is, interestingly, different from the other patterns. Ninety-six of these deaths, almost three-quarters, occurred in and around Belfast. Some of the remainder took place in the parts of north Armagh and east Tyrone in which many other deaths occurred; however, there are other clusters in parts of south Antrim and east Down where there were few other deaths. This suggests that these killings were occurring not so much in contested areas but in areas with large Protestant majorities and lower

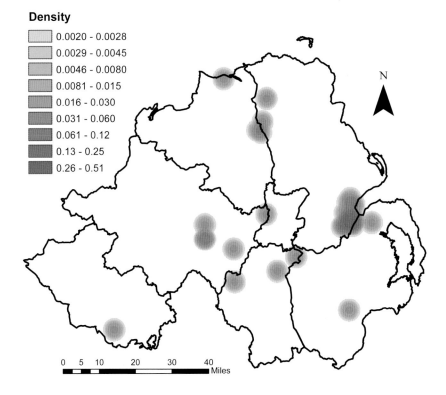

Density

	0.0020 - 0.0028
	0.0029 - 0.0045
	0.0046 - 0.0080
	0.0081 - 0.015
	0.016 - 0.030
	0.031 - 0.060
	0.061 - 0.12
	0.13 - 0.25
	0.26 - 0.51

N

0   5   10   20   30   40
Miles

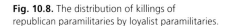

Fig. 10.8. The distribution of killings of republican paramilitaries by loyalist paramilitaries.

levels of violence. They may have been the consequence of feuding or of punishment killings.

Figures 10.10 and 10.11, respectively, show the republican paramilitaries and Catholic civilians killed by the security forces; these deaths represent 85 percent of the security forces' victims. Apart from the clusters in Belfast, Londonderry/Derry, and Strabane, the pattern of killings is quite dispersed, although County Londonderry and Antrim, away from the two urban centers, had very low levels of these deaths. Killings of republicans were more concentrated in Armagh, both north and south, and in east Tyrone as well as the three urban areas described above. Security force killings of Protestant civilians and loyalist paramilitaries have not been mapped because of their low numbers.

The importance of Northern Ireland's religious demography in shaping these patterns has already been hinted at. Figure 10.12a makes this explicit by summarizing the religious demography of the grid squares in which killings took place. The organizations responsible for the killings are depicted in figures 10.12b, 10.12c, and 10.12d. The background religious demography is calculated using the total number of Catholics recorded in the censuses of 1971, 1991, and 2001 divided by the total population in those years. Three clear peaks are discernible from the graphs. The largest is in what will be termed "Catholic areas," those areas where Catholics made up 80 to 90 percent of the population. The next is in "Protestant areas," where Catholics made up less than 10 percent of the population. The third is "mixed areas," in which violence peaks in grid squares that were 50 to 60 percent Catholic.

Grid squares that were 80 to 90 percent Catholic contained almost 30 percent of the total Catholic population, whereas those that had 90 percent and above Catholic populations or 70 to 80 percent both contained less than 10 percent. Twenty percent of all the killings that took place in the

Fig. 10.9. The distribution of killings of civilians by loyalist paramilitaries: (a) Catholics and (b) Protestants.

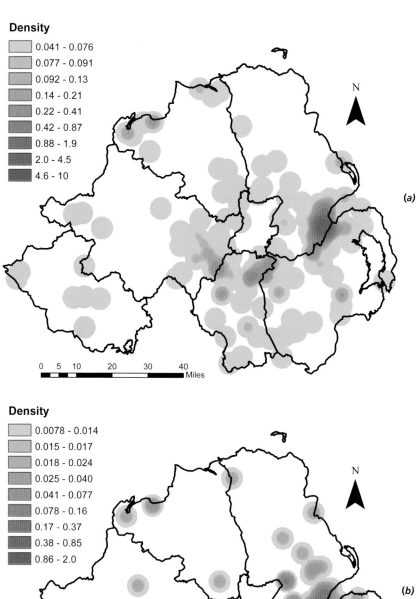

Density

- 0.041 - 0.076
- 0.077 - 0.091
- 0.092 - 0.13
- 0.14 - 0.21
- 0.22 - 0.41
- 0.42 - 0.87
- 0.88 - 1.9
- 2.0 - 4.5
- 4.6 - 10

N

(a)

0 5 10   20   30   40
Miles

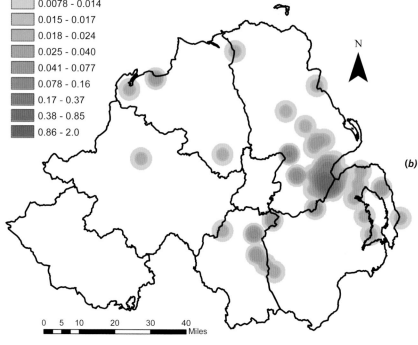

Density

- 0.0078 - 0.014
- 0.015 - 0.017
- 0.018 - 0.024
- 0.025 - 0.040
- 0.041 - 0.077
- 0.078 - 0.16
- 0.17 - 0.37
- 0.38 - 0.85
- 0.86 - 2.0

N

(b)

0 5 10   20   30   40
Miles

Troubles took place in the 80 to 90 percent Catholic decile. Of the victims of violence in this decile, the largest number was from the security forces; however, Catholic civilians and republican paramilitaries also died in significant numbers in these areas. This suggests that these small areas, covering less than 5 percent of Northern Ireland's land surface, were the scene of some of the most intense violence, mostly associated with the conflict between the security forces and republican paramilitaries, with Catholic civilians also being killed in large numbers. This is borne out by figures

Density

- 0.0059 - 0.011
- 0.012 - 0.013
- 0.014 - 0.018
- 0.019 - 0.030
- 0.031 - 0.058
- 0.059 - 0.12
- 0.13 - 0.28
- 0.29 - 0.64
- 0.65 - 1.5

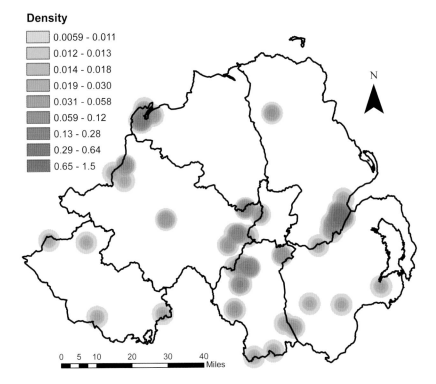

0   5   10   20   30   40
Miles

Fig. 10.10. The distribution of killings of republican paramilitaries by the security forces.

10.12b and 10.12d, which show that over 25 percent of the killings by both republican paramilitaries and the security forces took place in these areas. By contrast, these areas only contain 7.7 percent of loyalist killings, suggesting that, despite the large numbers of potential victims, loyalists tended to avoid these areas.

Interestingly, the next highest number of killings took place at the opposite end of the religious spectrum—in the decile that was less than 10 percent Catholic. Grid squares with this demographic contained 28.8 per-

Density

- 0.010 - 0.019
- 0.020 - 0.023
- 0.024 - 0.032
- 0.033 - 0.054
- 0.055 - 0.10
- 0.11 - 0.22
- 0.23 - 0.50
- 0.51 - 1.1
- 1.2 - 2.7

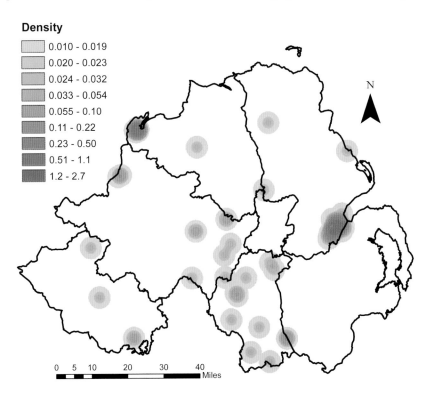

0   5   10   20   30   40
Miles

Fig. 10.11. The distribution of killings of Catholic civilians by the security forces.

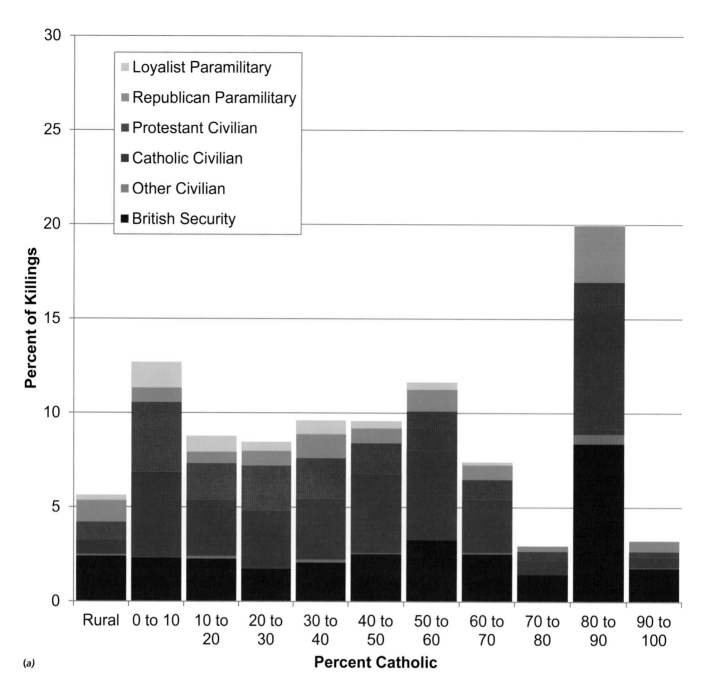

Percent Catholic

(a)

Fig. 10.12a. The distribution of killings by the religious background of the 1 km grid square in which the killing took place: (a) all killings, (b) killings by republican paramilitaries. The legend refers to the victim of the killing. Percent Catholic calculated as the average over 1971, 1991, and 2001. "Rural" refers to places that have a total population over the three decades of fewer than fifty.

cent of the total population and 42.3 percent of the non-Catholic population. These Protestant areas saw 12.7 percent of the total killings; however, the characteristics of the conflict in these areas were very different. They were the focus of much loyalist violence, containing 23.6 percent of their killings, compared to only 8 percent of republican killings and 5.8 percent of security force killings. The distribution of victims is also interesting. Of the 444 people killed in these areas, 158 were Catholic civilians, 129 were Protestant civilians, 81 were from the security forces, 48 were loyalists, and 27 were republicans. This shows that there was a very different pattern of conflict in Protestant areas compared to Catholic ones. Figure 10.12c shows that in the 0 to 10 percent decile, loyalists primarily killed Catholic civilians as well as Protestant civilians and other loyalist paramilitaries. By contrast, figure 10.12b shows that republicans largely killed security forces and Protestant civilians, while figure 10.12d shows that the security forces were most

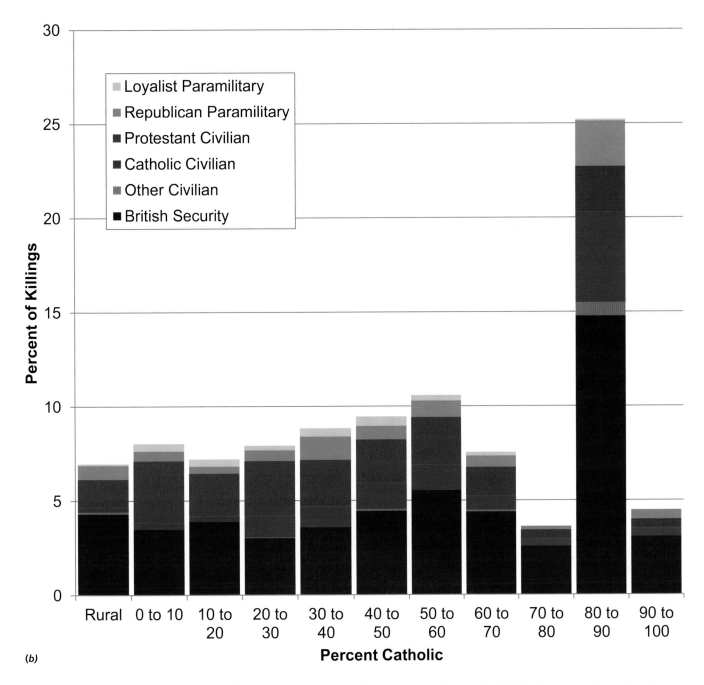

(b)

Fig. 10.12b. (Continued on the next page)

successful in avoiding civilian casualties in these areas, with 39 percent of their victims being republicans and a further 26 percent being loyalists.

Three dynamics of violence can thus be identified in these areas. The first is similar to that found in Catholic areas. Just as republicans killed the largest numbers of their victims in what might be termed their own areas, so too with loyalists, who killed the largest numbers of their victims in Protestant areas. Beyond this, however, there was a second dynamic that was very different in Protestant areas compared to Catholic ones. Loyalist victims in Protestant areas were primarily civilians rather than members of the security forces. These civilians came from both religions, although it is important to remember that there would have been far fewer Catholics in these areas, so their death rate would have been proportionately far higher than that of Protestants. Unlike Catholic areas, however, where the security forces caused large numbers of civilian deaths, in Protestant areas the secu-

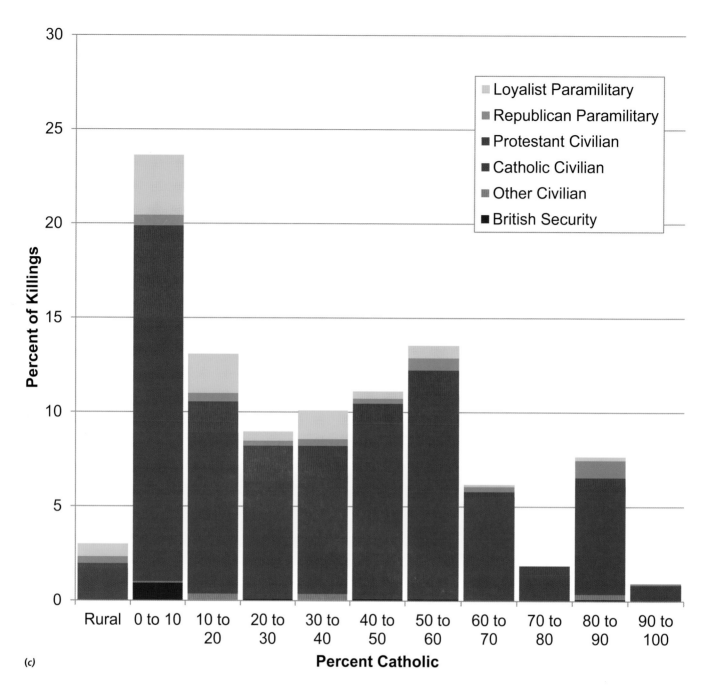

30

25

20

15

10

5

0

Percent of Killings

- Loyalist Paramilitary
- Republican Paramilitary
- Protestant Civilian
- Catholic Civilian
- Other Civilian
- British Security

Rural | 0 to 10 | 10 to 20 | 20 to 30 | 30 to 40 | 40 to 50 | 50 to 60 | 60 to 70 | 70 to 80 | 80 to 90 | 90 to 100

(c)

Percent Catholic

Fig. 10.12c and 10.12d. The distribution of killings by the religious background of the 1 km grid square in which the killing took place (c) killings by loyalist paramilitaries, and (d) killings by the security forces. The legend refers to the victim of the killing. Percent Catholic calculated as the average over 1971, 1991, and 2001. "Rural" refers to places that have a total population over the three decades of fewer than fifty.

rity forces killed relatively few people, and many of those killed were in fact republicans. Republicans, in turn, tended to kill members of the security forces or loyalists. This points to the third dynamic: the continuation of the conflict between republicans and the security forces even in Protestant areas, with frequently lethal consequences for each other and for Protestant civilians.

The third highest concentration of violence took place in "mixed areas," particularly those that were 50 to 60 percent Catholic. All three groups of protagonists killed their second largest number of victims in areas with this demographic despite the fact that they only contained 5.3 percent of the total population. For loyalists, these victims were overwhelmingly Catholic civilians. For republicans, they were the security forces and Protestant civilians. For the security forces, they were Catholic civilians and republican

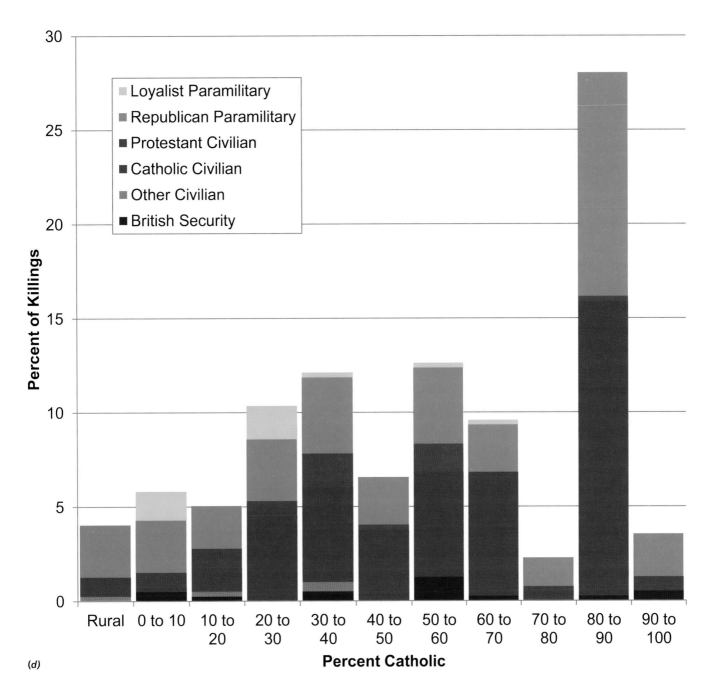

(d)

Percent Catholic

paramilitaries. Here, therefore, we have both major types of conflict going on: the security forces against the republicans and vice versa, with many civilian casualties, and loyalists killing Catholic civilians without coming into major conflict with the security forces. Other smaller elements of the conflict, particularly paramilitaries killing other members of their own community, either civilians or paramilitaries, are much less common in these mixed areas.

These figures bear out that there were two major types of conflict in Northern Ireland during the Troubles as well as a number of more minor ones. Within these categories there was also a highly territorialized pattern of violence linked to entrenched spatioreligious patterns and the contested nature of political territoriality. There were also social patterns to the violence concerned with the socioeconomic composition of the areas in which violence occurred and where both urban deprivation and rurality played their parts in defining different patterns of conflict. These factors will be returned to in the next chapter.

It is difficult to be too prescriptive about the causes of these geographies; however, the pattern of Catholic civilian killings by loyalists does fit closely with ideas about political territoriality generally. More specifically, it reflects the fact that loyalist paramilitaries chose to focus their aggression toward particular Catholic enclaves in north rather than west Belfast, suggesting that geographical imperatives had an important role to play in influencing the behavior of loyalist paramilitaries.[16] At the same time, deaths within Catholic areas suggest a conflict in which republicans were trying to force the security forces—the representatives of U.K. governance in Northern Ireland—out of these areas, while the security forces fought to maintain their influence within them.[17]

Conclusions

The maps show that conflict-related deaths during the Troubles were highly geographically concentrated. Most of Northern Ireland did not experience fatal violence directly, although the indirect effects in terms of political tension, fear, and knowledge of the victims were more widely spread. The spatial concentration of violence, however, meant that some neighborhoods suffered the direct impacts of the conflict disproportionately both at the time and after it was over. This legacy of violence, evidenced by high rates of antidepressant use, social deprivation, out-migration, and communal malaise, has blighted parts of central Belfast and parts of the city of Londonderry/Derry. The maps also show the multifaceted nature of the conflict, with differing geographies of death for killings committed by republicans, loyalists, and security forces. Territoriality is clearly important as something to control both externally at the expense of the "other group," particularly loyalists killing Catholics in Protestant areas, and internally by killing one's "own" group. Overall, the maps indicate that the Troubles were not a two-sided conflict but had three parties with varying relationships and also external and internal dynamics.

Belfast through the Troubles: Socioeconomic Change, Segregation, and Violence   11

The city of Belfast provides an illustration of much of the division and interdependence that have taken place in Ireland over the last two centuries. Belfast's success as an industrial city did much to separate the economy and outlook of the Protestant northeast of Ulster from the rest of Ireland. That success had much to do with developing strong links between this part of Ireland and Britain, but it also brought many Catholics from elsewhere in Ireland to the city in search of jobs. This left Belfast with complex spatioreligious patterns that, when the Troubles started in the late 1960s, were particularly contested, resulting, as chapter 10 identified, in the city being the focus for much of the ensuing violence. Over the same period, as described in chapter 9, Belfast went through a period of rapid deindustrialization. Against this background, this chapter first examines the evolving religious geography of the city and related developments in residential segregation. Second, it considers how Belfast has changed socially and economically over the thirty-year period between 1971 and 2001. Third, it looks at the patterns of violence within the city. Finally, the chapter draws these themes together to show how changing spatioreligious patterns and levels of residential segregation are related to wider socioeconomic trends, thereby trying to set Belfast's experience within the broader context of urban change as observed in other societies.

Defining the boundaries of Belfast is not as easy as it might at first appear. There is no government authority that covers the whole metropolitan area or conurbation, and, indeed, the drawing of local government boundaries and decisions about the numbers and shapes of these units are highly controversial. The fifty-one electoral wards that make up the Belfast District Council provide a very tight definition of Belfast that excludes neighborhoods that should be considered to be part of "the city." Analyses that focus on the Belfast District Council therefore only deal with the core of the city, ignoring most of the urban fringe and suburbs. Because of this, we use the Belfast Urban Area (BUA) to delimit the city. Although this was originally developed for use in the 1991 census—and was not used in 1971 and 2001—it is appropriate because it covers most of the built-up area that can be considered to be part of late twentieth-century Belfast and escapes the geographical confines of the Belfast District Council area. The BUA incorporates parts of Lisburn, Newtownabbey, Castlereagh, and north Down, thereby allowing us to examine changes in the urban core and in the outer suburbs together. The BUA is shown in figure 11.1, which also locates many of the places described below.

In this chapter we use the 100 m grid square product for 1971 to 2001 for our analysis. This permits demographic, social, and economic profiles

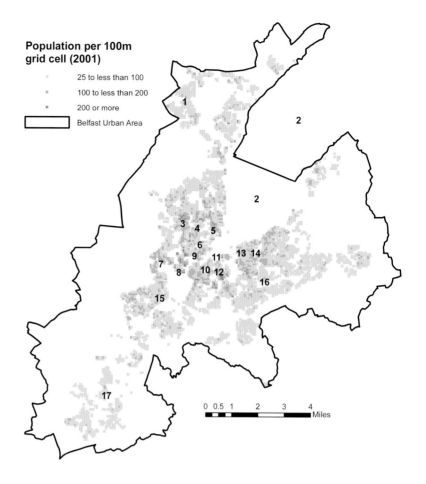

Fig 11.1. The Belfast Urban Area. Unshaded grid squares have populations of fewer than twenty-five. Legend: Andersonstown (15); Ardoyne (3); Ballymurphy (7); Belfast Lough (2); Castlereagh (16); city center/city hall (11); Donegall Pass (12); Falls (Lower) (9); Falls (Upper) (8); Lisburn (17); Lower Newtownards Road (14); New Lodge (5); Newtownabbey (1); Oldpark (4); Sandy Row (10); Shankill (6); Short Strand (13).

to be constructed at a very small spatial scale, equivalent to two American football fields. This finely meshed analysis is useful in urban areas where populations are dense and where large social or demographic changes can occur over short distances. The use of these data also allows detailed comparisons of how places changed through time.

Religious and Residential Segregation: The "Exceptional City"

Figure 11.2 shows Belfast's religious geographies in 1971 and 2001. Moving clockwise from the south shore of Belfast Lough, the basic pattern is of Protestants in the east and a relatively mixed area in the more middle-class south. West of this, the Catholic heartland runs southwest from the city center through the Falls to Andersonstown. North of this is the Protestant Shankill, which runs due west from the city center. North of this again is a more complex part of the city that comprises both Catholic and Protestant areas, including the Ardoyne and New Lodge. Moving north away from the city center, the city again becomes dominated by Protestants.

Figure 11.2 suggests that by 2001 Belfast was more Catholic and more segregated than it was in 1971, and this is backed up by the figures. In 1971 Belfast's population was 24.8 percent Catholic. This rose to 31.6 percent in 1991 and to 35.2 percent or 39.5 percent by religious affiliation and community background, respectively, in 2001. While these observations are simple, the path by which Belfast reached this endpoint in 2001 is complicated, involving absolute and relative shifts in population numbers and

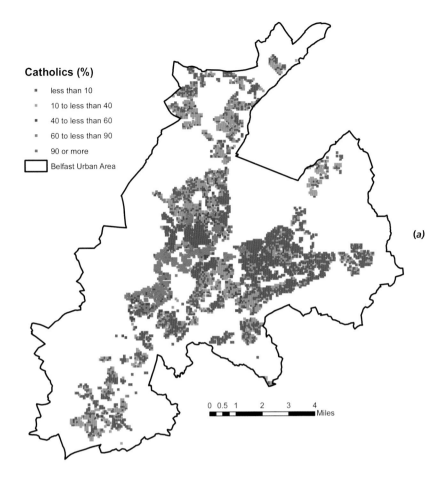

Catholics (%)

- less than 10
- 10 to less than 40
- 40 to less than 60
- 60 to less than 90
- 90 or more
- Belfast Urban Area

0 0.5 1   2   3   4
Miles

(a)

Fig. 11.2. Belfast's religious geographies, (*a*) 1971 and (*b*) 2001.

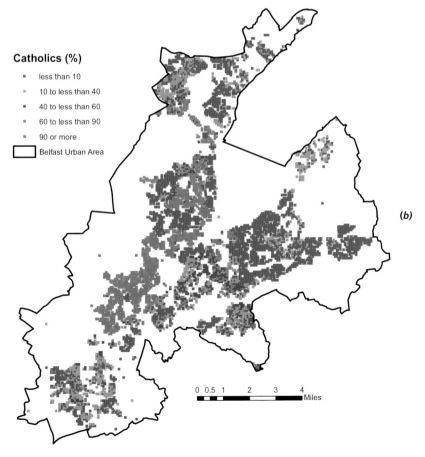

Catholics (%)

- less than 10
- 10 to less than 40
- 40 to less than 60
- 60 to less than 90
- 90 or more
- Belfast Urban Area

0 0.5 1   2   3   4
Miles

(b)

Fig. 11.3. Population change in Belfast, 1971–2001, for (a) total population, (b) Catholics, and (c) Protestants.

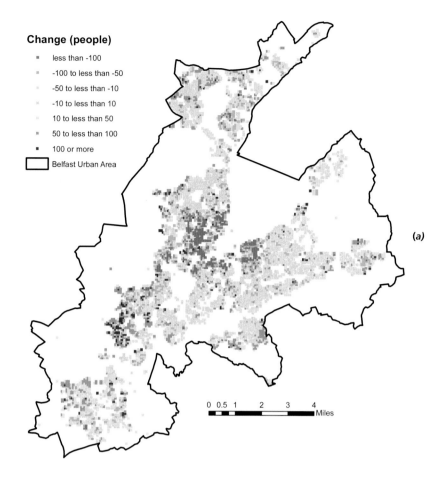

Change (people)
- less than -100
- -100 to less than -50
- -50 to less than -10
- -10 to less than 10
- 10 to less than 50
- 50 to less than 100
- 100 or more
- Belfast Urban Area

(a)

0 0.5 1 2 3 4 Miles

locations. The overall context, shown in figure 11.3a, is set by the decline in the population of Belfast as a whole, with the highest population declines in the urban core. In 1971 Belfast had a population of 485,000, which fell by 26.6 percent to 356,000 in 1991 before recovering to 437,000 in 2001. The growth in residential segregation and the proportionate Catholic increase happened against this backdrop of general population decline. As the maps in figure 11.3b and 11.3c show, population decrease affected both Catholics and Protestants but had unequal impacts and unequal outcomes. Figure 11.3b indicates that the numbers of Catholics fell in the inner city but grew along axes southwest and north of the city center, especially in the southwest of Belfast. Overall, Catholic percentages in Belfast rose by 27.9 percent or 43.4 percent between 1971 and 2001 using religious affiliation and community background, respectively. Over the same period, the Protestant population fell by 35.3 percent or 22.3 percent, respectively. As figure 11.3c shows, this decrease was far more geographically widespread than that experienced by the Catholic community. Neighborhoods that saw the largest falls include the Shankill in the west, Lower Newtownards Road immediately east of the city center, and Donegall Pass and Sandy Row immediately to its south. In 1971 these had been densely populated areas with strong working-class traditions associated with Belfast's heavy industries. This is in some ways similar to the Catholic pattern in that it shows loss from inner-city areas; however, the geography of Protestant population increase was very different. Increases in the Protestant population were concentrated on the outer fringe of the urban area, whereas the Catholic distribution

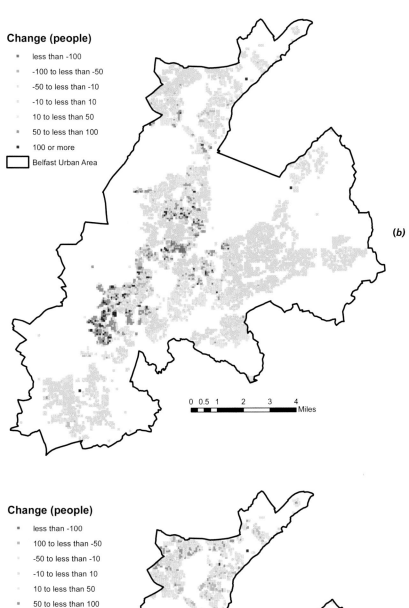

Change (people)

- ■ less than -100
- ■ -100 to less than -50
- ■ -50 to less than -10
- ■ -10 to less than 10
- ■ 10 to less than 50
- ■ 50 to less than 100
- ■ 100 or more
- ☐ Belfast Urban Area

(b)

0 0.5 1   2   3   4
Miles

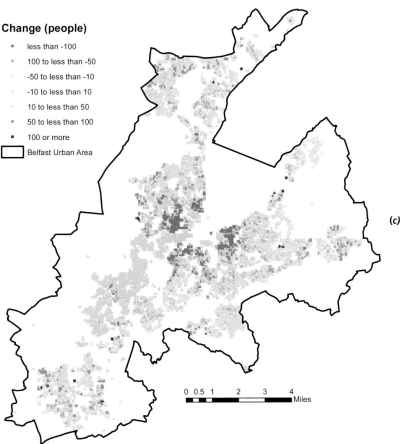

Change (people)

- ■ less than -100
- ■ 100 to less than -50
- ■ -50 to less than -10
- ■ -10 to less than 10
- ■ 10 to less than 50
- ■ 50 to less than 100
- ■ 100 or more
- ☐ Belfast Urban Area

(c)

0 0.5 1   2   3   4
Miles

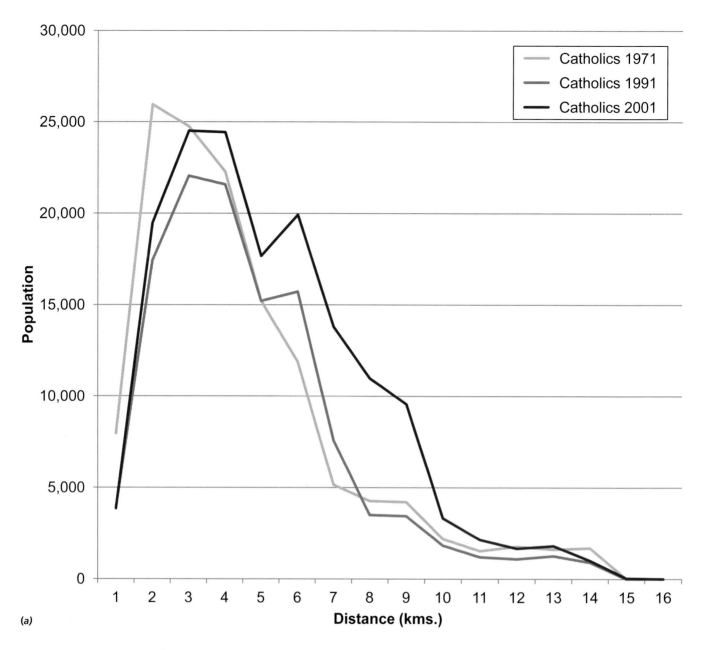

(a)

Fig. 11.4. Population by distance from city hall among (a) Catholics and (b) Protestants. Distances are in kilometers.

was more centrally located. Anecdotal evidence suggests that figure 11.3c, if anything, underplays the degree of Protestant decentralization from central Belfast, since many Protestants appear to have left the city altogether to move to nearby towns such as Ballyclare, Comber, and Newtownards.

Figure 11.4 presents an alternative way of visualizing the changing distribution of Belfast's religious communities. The graphs show population numbers by religious denomination in 1971, 1991, and 2001 for Catholics and Protestants in kilometer distance bands measured from city hall. In effect, they are transects of the city population, their contours showing the spatial concentration of the population. Key features to note are that the Catholic population did not change much in terms of its absolute numbers between 1971 and 2001. Its spatial structure also remained fairly stable, although there is some evidence that numbers farther away from the city center had grown by 2001. The Protestant population, however, is far more dynamic. There are large falls in Protestant numbers in the distance bands between 2 km and 6 km from city hall between 1971 and 1991, while num-

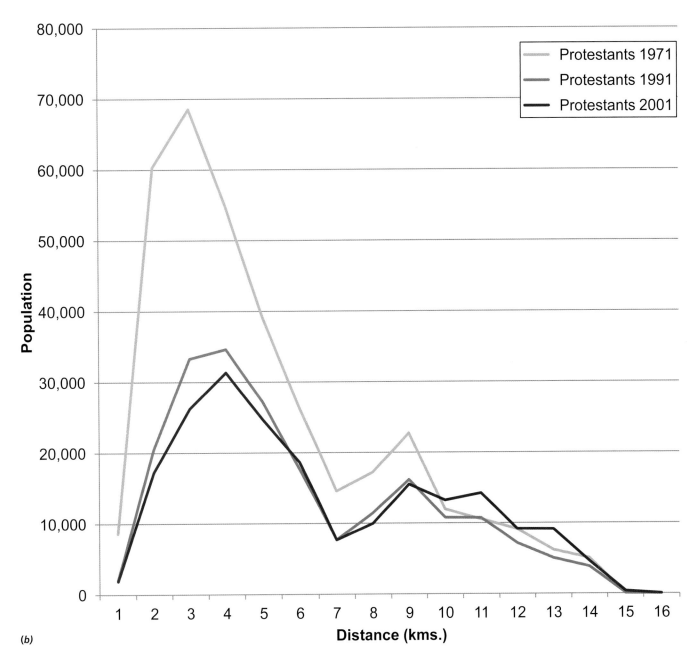

(b)

bers farther away remain more stable. There was far less Protestant change between 1991 and 2001. The effect of these changes has been to make the areas near the city center, and indeed Belfast as a whole, proportionately more Catholic through time, and, as the Protestant inner-city population has fallen, Protestants have become proportionately more suburbanized than Catholics.

The outcome of these differential patterns of population change has been to lead to increases in indexes of segregation since 1971. The overall values of the segregation indexes, however, increased most between 1971 and 1991 and peaked in most cases in 1991 before remaining at the same level or decreasing slightly in 2001. The dissimilarity index rose from 68.3 percent in 1971 to around 80 percent in both 1991 and 2001. The Catholic isolation index rose from 57.5 percent in 1971, 32.7 percentage points above expected, to around 75 percent in 1991 and 2001, which is around forty points above expected. The Protestant isolation index fell from 82.6 percent in 1971 to around 70 percent in 1991 and 2001 but rose from being 19.1 points above expected in 1971 to 23.9 points in 2001. This accords well with the evidence from the graphs: Protestant decrease and proportionate Catholic increase were greatest between 1971 and 1991 and led to greater segregation, especially for Catholics.

There are limits to how much cross-sectional census data can help our understanding of the demographic processes that led to these changes. For example, even the extent to which these changes were due to migration or differentials in births and deaths is unclear. It is even harder to be sure about the causes of these processes; however, it seems likely that Protestant out-migration from Belfast in the 1970s and 1980s was an important element driving population change. Moreover, public housing policy, combined with trends in the private sector, seems also to have played a significant element in concentrating the Catholic population in the southwest of the city. Further comments on process and some views on causation are developed later in the chapter.

Industrial Decline: Typicality of Other Dimensions of Change

The emphasis on residential segregation and demographic change by community background sets Belfast apart from Irish cities south of the border and other cities in the U.K. Despite this, in other ways Belfast is similar to other U.K. and Irish cities. As has already been described, its nineteenth-century industrial base had much in common with the industrial cities of northern England and western Scotland. Changes in housing tenure also mirror the British experience: from 1971 to 2001 there was an expansion in the proportion of the housing stock that was owner occupied, and, as figure 11.5 shows, there was a clear geography to this expansion. In 1971 there were large numbers of estates where almost every house was socially rented, in other words, owned by local government and rented to tenants often on the basis of social need. By 2001, although "right-to-buy" legislation had reduced social renting dramatically, it remained prevalent in inner-city and western areas, although the exclusively rented areas from 1971 had all but disappeared.

Belfast has also seen a decline in employment rates in city center communities, while employment has increased on the periphery, as illustrated by figure 11.6. The reasons for this change are many and complex. The higher employment rates in the periphery are perhaps the result of more mobile, skilled, and educated people leaving the city center for homes in desirable suburban locations. This selective out-migration tells only part of the story. The loss of jobs in the heavy industries that employed many workers before the 1970s has also contributed to the economic difficulties of the inner-city population, as many people in these areas were dependent on the manufacturing sector. As with many other cities, this sector experienced a gradual but long decline after World War II, with the Harland and Wolff shipyards laying off their last workers in 2000.[1] It is difficult to map occupational change over time because the categories used change between censuses. Despite this difficulty, there is little doubt that Belfast has seen an expansion of service employment since 1971, with the growth of call centers, retailing, and the public sector all contributing to this shift.

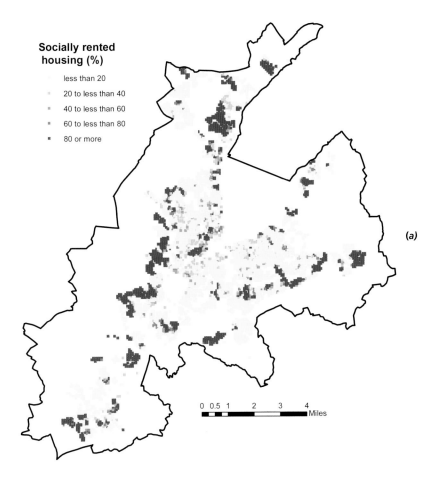

**Socially rented
housing (%)**

less than 20

20 to less than 40

40 to less than 60

60 to less than 80

80 or more

Fig. 11.5. Socially rented housing in
Belfast in (a) 1971 and (b) 2001.

(a)

0 0.5 1   2   3   4
Miles

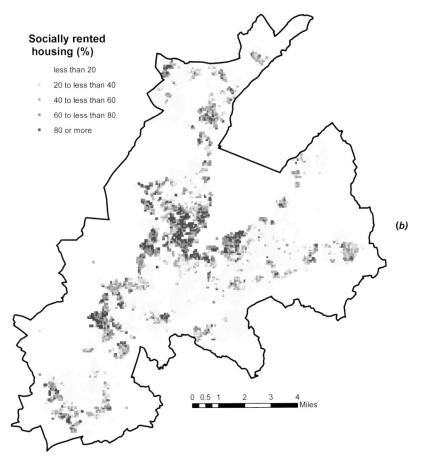

**Socially rented
housing (%)**

less than 20

20 to less than 40

40 to less than 60

60 to less than 80

80 or more

(b)

0 0.5 1   2   3   4
Miles

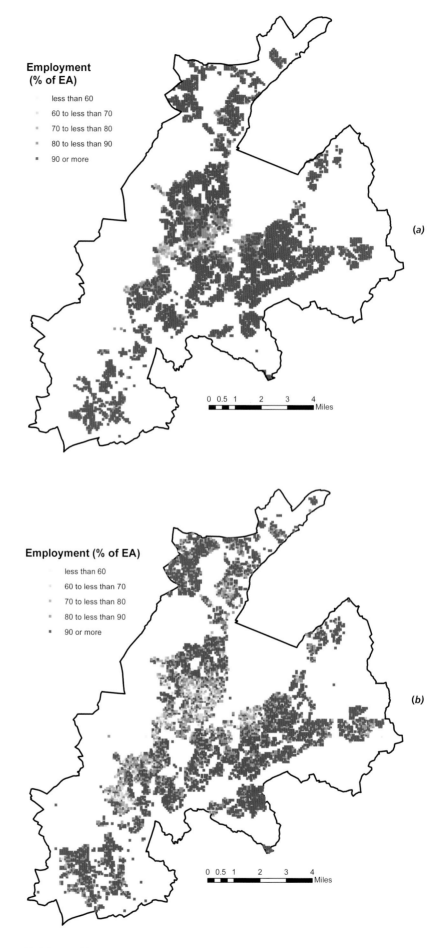

Fig. 11.6. Employment as a percentage of the economically active population in Belfast in (a) 1971 and (b) 2001.

As with Northern Ireland as a whole, the residents of Belfast have, on average, become more affluent. Car ownership, for example, as shown in figure 11.7, was much more prevalent in 2001 than in 1971, which shows that the proportion of households with no cars fell through time. Finally, figure 11.8 illustrates how the average size of households has gotten smaller through time in Belfast, with, by 2001, most of the urban area having smaller average households than in 1971.

In experiencing these socioeconomic and demographic changes since the early 1970s, Belfast fits closely into the general urban patterns established elsewhere. Population loss, especially from the urban core, is part and parcel of these patterns, and deindustrialization and counterurbanization could just as easily apply to Belfast as they could to Manchester, Sheffield, or Detroit. More generally, these aspects of change are all dimensions of general descriptions of restructuring that have been theorized to address these and similar developments throughout Europe and North America.[2] Contemporary Belfast, in comparison with 1971, could very plausibly now be described as a postindustrial or post-Fordist city. Changes in the religious demography and levels of residential segregation need to be understood in the context of these macroeconomic and social trends.

A Conflict within a Conflict: Violence in Belfast

The previous chapter used the Sutton database of deaths to describe patterns of violence during the Troubles. It identified that the Troubles consisted of a number of separate but related conflicts, particularly one in which republican paramilitaries killed security forces and vice versa, with both also killing civilians, and a second in which loyalists primarily killed Catholic civilians. It also identified that patterns of violence had clear geographies that were determined in part by the underlying demographic and socioeconomic geographies. Within these geographies, different types of conflict, as measured by perpetrator and victim, also had clearly different geographies. While chapter 10 identified Belfast as a major center for many forms of violence, because it was looking at all of Northern Ireland using 1 km grid square data, the chapter was unable to make many statements about what happened locally within Belfast. This section analyzes patterns of violent death in Belfast during the Troubles in more detail in order to explore, as far as we can, both how background religious and socioeconomic geographies affected the patterns of violence and how, in turn, patterns of violence may have affected religious and socioeconomic geographies.

The Sutton database records 1,633 Troubles-related deaths in the BUA, slightly under half of all the deaths recorded during the Troubles. A density-smoothed map of these deaths is shown in figure 11.9, which also shows the locations of Belfast's main peace lines. The major peace line that divides the Catholic Falls from the Protestant Shankill runs west in a near straight line from the city center, splitting west Belfast into two. The other lines are often more complex, particularly in north Belfast, reflecting the more complex religious geographies in these areas. The map shows that while many of the populated parts of the BUA were affected by violence to one extent or another, most violent deaths were concentrated in a relatively small area in the city center and west and north of it. These can be

Fig. 11.7. Households lacking a car in Belfast in (*a*) 1971 and (*b*) 2001.

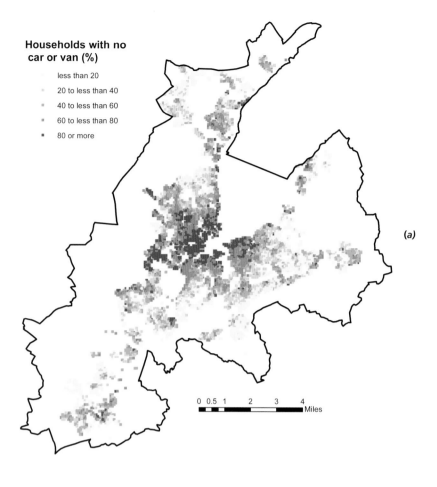

Households with no car or van (%)

less than 20
20 to less than 40
40 to less than 60
60 to less than 80
80 or more

0 0.5 1 2 3 4
Miles

(*a*)

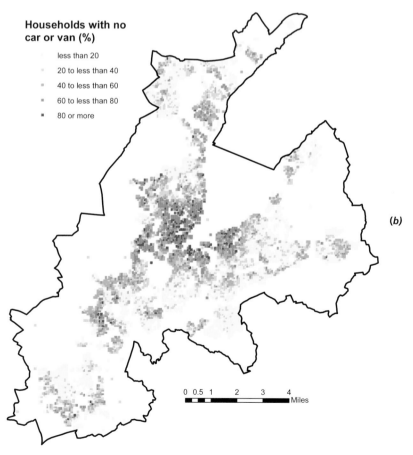

Households with no car or van (%)

less than 20
20 to less than 40
40 to less than 60
60 to less than 80
80 or more

0 0.5 1 2 3 4
Miles

(*b*)

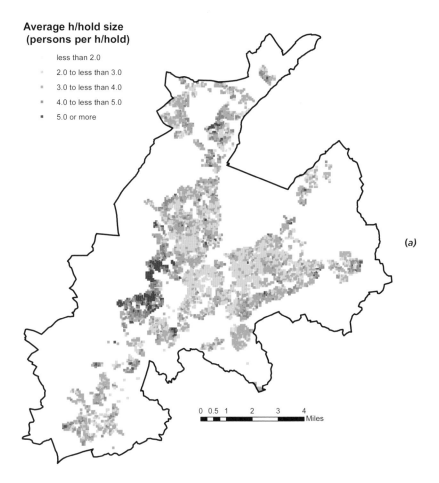

**Average h/hold size
(persons per h/hold)**

　　less than 2.0
▪　2.0 to less than 3.0
▪　3.0 to less than 4.0
▪　4.0 to less than 5.0
▪　5.0 or more

(a)

0　0.5　1　　2　　3　　4
Miles

Fig. 11.8. Average household size in
Belfast in (a) 1971 and (b) 2001.

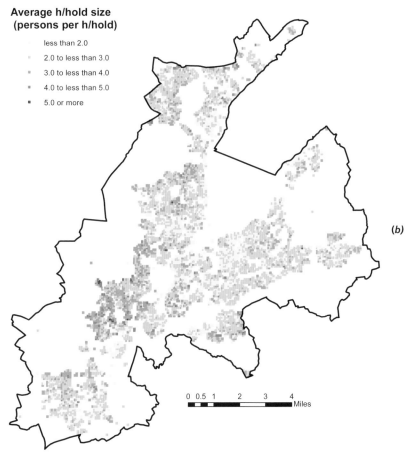

**Average h/hold size
(persons per h/hold)**

　　less than 2.0
▪　2.0 to less than 3.0
▪　3.0 to less than 4.0
▪　4.0 to less than 5.0
▪　5.0 or more

(b)

0　0.5　1　　2　　3　　4
Miles

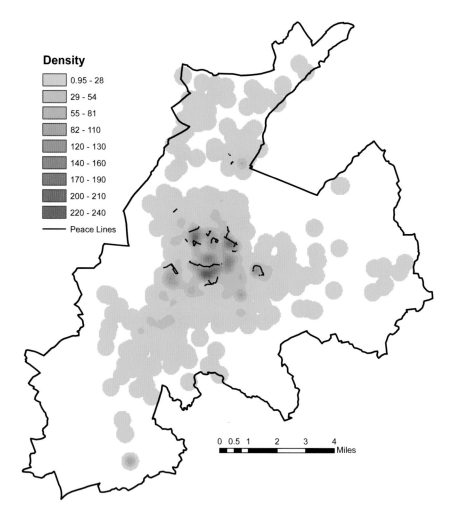

Density

- 0.95 - 28
- 29 - 54
- 55 - 81
- 82 - 110
- 120 - 130
- 140 - 160
- 170 - 190
- 200 - 210
- 220 - 240
- —— Peace Lines

0 0.5 1 2 3 4
Miles

further subdivided into seven main clusters: the first is the city center at the eastern end of the Falls/Shankill peace line; next is the Lower Falls, west of the city center and south of the Falls/Shankill peace line; west again is a double cluster that consists of the Upper Falls and Ballymurphy; north of the Lower Falls there is a cluster in the Shankill; north and west of this there is another double cluster hemmed in by five short peace lines that includes the Ardoyne and Oldpark; east of this is a cluster in New Lodge directly to the north of the city center; finally, east of the city center is the Lower Newtownards Road cluster, which lies in and around the Short Strand, a Catholic enclave almost completely enclosed by a single peace line. The definitions and extents of these clusters are, to a certain extent, arbitrary, but they nevertheless provide a useful framework to explore patterns of violence. Other, smaller clusters can also be identified, but it is also worth noting that there are large areas, mainly farther away from the city center, that had little or no fatal violence over the entire period.

These seven main clusters provide a useful framework for further analyzing patterns of violence in Belfast. As table 11.1 shows, there were 885 killings in the seven clusters, 52.4 percent of the total killings in the city and a quarter of the entire total during the Troubles. These took place in a combined area of only 10 km²; thus, over half of the killings within the BUA took place in only 3.8 percent of the BUA's total area and a tiny fraction of the whole of Northern Ireland. These areas contained only 23.5 per-

Table 11.1. The religious background of the major clusters of Troubles-related deaths, 1971 and 2001

Cluster	Deaths	Area (km²)	% of deaths	% of pop. 1971	% of pop. 2001	% of Cath. 1971	% of Prot. 1971	Catholic 1971 (%)	Protestant 1971 (%)	Catholic 2001 (%)	Protestant 2001 (%)
City center	107	1.2	6.6	0.5	0.4	1.5	0.1	70.1	13.7	62.2	36.7
Lower Falls	192	1.5	11.8	3.8	1.9	11.5	0.4	75.0	6.6	96.6	2.7
Upper Falls and Ballymurphy	116	1.7	7.1	3.3	2.6	10.5	0.2	78.3	3.1	95.3	2.8
Shankill	100	1.3	6.1	3.9	1.6	0.2	5.7	1.5	95.0	4.3	93.0
Ardoyne and Oldpark	199	2.2	12.2	6.0	3.8	7.9	5.1	33.0	55.1	66.1	32.4
New Lodge	107	1.3	6.6	2.9	1.8	5.5	1.7	46.8	38.1	80.5	18.1
Lower Newtownards Road	64	1.2	3.9	3.1	1.5	2.6	3.4	20.7	71.3	37.2	59.2
All clusters	885	10.3	54.2	23.5	13.6	39.7	16.5	24.8	64.9	39.5	56.0
BUA total	1,633	272.2	100	100	100	100	100	42.0	45.5	67.0	31.2

Note: Census data are taken from the 100 m grid square product. The 2001 numbers of Catholics and Protestants are by community background.

cent of Belfast's population in 1971, dropping to 13.5 percent by 2001. This population was not divided evenly by religion. In 1971 almost 40 percent of the city's Catholics lived in these areas compared to only 16.5 percent of Protestants. By 2001 these figures had dropped to 23.1 percent and 7.6 percent, respectively, as measured by community background. This means that overall the Catholics of Belfast were far more exposed to violence than the Protestants. However, when clusters are explored individually, there are significant differences in their religious demography. Figures are provided for the city center, but these are perhaps misleading, as violence in this area was probably more related to its commercial functions than to its residential characteristics. The Lower Falls as well as the Upper Falls and Ballymurphy clusters had large Catholic majorities of around 75 percent in 1971, rising to over 95 percent by community background in 2001.[3] The Shankill had very much the opposite demographic. This cluster was 95 percent Protestant in 1971, a figure that had only declined slightly by 2001. The north Belfast clusters reveal a very different pattern: they were less than 50 percent Catholic in 1971 but had significant Catholic majorities by 2001. Finally, the Lower Newtownards Road cluster showed a similar pattern of proportional Catholic increase, although Catholics did not become a majority in this area.

There was thus a pattern of violent areas that had Catholic and mixed populations becoming proportionately more Catholic, while the Protestant Shankill remained staunchly Protestant. It should be noted that this pattern was not a result of population growth. Almost all these areas suffered population losses of between one-third and two-thirds over the thirty-year period. In almost every case this loss applied to both religious groups, but Protestant loss was generally greater than Catholic. The exception to this is the Catholic population of the Ardoyne and Oldpark cluster, which went up by 14 percent.

The differences in the religious demography of areas in which violence was prevalent perhaps inevitably led to differences in who perpetrated the violence and who their victims were. The Lower Falls and the Upper Falls and Ballymurphy clusters show evidence primarily of conflict between republicans and security forces. In the Lower Falls republicans killed 116 of the 192 people who died, with almost half being from the security forces and a further 35 being civilians, of whom 22 were Catholics. The security forces were also active in the Lower Falls, killing forty people, of whom Catholic civilians outnumbered republicans by two to one. This points to the conflict between republicans and the security forces, with many civilian casualties. The second type of conflict, in which loyalists targeted Catholic civilians, can also be shown to have occurred here: almost all of the thirty-three victims of loyalist violence in these areas were Catholic civilians. Finally, republican on republican violence killed a further twenty-one people. The Upper Falls and Ballymurphy cluster shows a similar pattern, although here loyalists killed fewer people.

In the Shankill the largest number of killings was perpetrated by loyalists, who killed forty-seven of the one hundred victims in this area. The loyalists' victims were divided between seventeen Catholic civilians, fourteen Protestant civilians, and thirteen other loyalists. However, they did not kill a single member of the security forces in this area, and, in turn, the security forces killed only one loyalist. Therefore, while there was widespread loyalist violence, it killed civilians from both sides and other loyalists, but there was little lethal conflict between loyalists and the security forces. Republican paramilitaries were also active, killing thirty-nine, of whom twenty-six were Protestant civilians and eight were loyalists, with only a few of the republicans' victims being from the security forces. The New Lodge cluster shows similar characteristics, with 107 killings, of which 52 were Catholic civilians killed by loyalists.

The Ardoyne and Oldpark area shows a more complex pattern of violence. This cluster showed the largest number of deaths, at 199, and there are large elements of both major types of conflict. Loyalists killed eighty-two, of whom sixty-one were Catholic civilians. Republicans killed almost as many, at seventy-nine, but thirty-eight of these were from the security forces, and twenty-seven were Protestant civilians. The security forces killed thirty-four, of whom fourteen were Catholic civilians and thirteen were republicans. While this provides evidence for both of the major types of conflict, there is a relative lack of paramilitaries from both sides killing members of their own community, either paramilitary or civilian, in this area. Another interesting finding is that the two areas in which loyalist violence against Catholic civilians was most pronounced showed major swings in their religious demography, from having a clear Protestant majority in the case of the Ardoyne and Oldpark and a mixed population in New Lodge to both having significant Catholic majorities.

The cluster around the Lower Newtownards Road provides a final contrast. There were sixty-four deaths in this area, of which the majority were civilians killed by paramilitaries from the opposite community. There is far less evidence of violence by and against the security forces here, with only four casualties and ten killings, of whom, unusually, seven were loyalists.

Although the account above suggests that the major clusters of violence had significant differences in terms of both their religious demography and the nature of the violence, there is one feature that they all had in common. Table 11.2 shows the socioeconomic background of the clusters, focusing on three variables commonly regarded as being associated with deprivation: socially rented housing, households lacking a car, and unemployment. In

Table 11.2. The socioeconomic background of the major clusters of Troubles-related deaths, 1971 and 2001

Cluster	Deaths	Area (km²)	% of deaths	Socially rented housing 1971 (%)	No car 1971 (%)	Unemployed 1971 (%)	Socially rented housing 2001 (%)	No car 2001 (%)	Unemployed 2001 (%)
City center	107	1.2	6.6	68.1	90.6	26.9	74.2	72.5	24.0
Lower Falls	192	1.5	11.8	49.9	87.0	18.5	60.4	70.8	23.8
Upper Falls and Ballymurphy	116	1.7	7.1	62.4	81.8	23.1	43.7	61.9	24.0
Shankill	100	1.3	6.1	14.3	80.2	11.3	59.9	68.3	19.8
Ardoyne and Oldpark	199	2.2	12.2	16.1	74.4	13.4	54.0	63.2	22.3
New Lodge	107	1.3	6.6	29.0	81.4	14.4	63.3	71.1	23.4
Lower Newtownards Road	64	1.2	3.9	14.2	81.9	13.0	69.6	67.3	23.1
All clusters	885	10.3	54.2	27.7	80.5	15.5	58.0	66.6	22.8
BUA total	1,633	272.2		31.9	56.7	8.3	29.3	37.4	10.0

Note: Census data are taken from the 100 m grid square product.

almost all cases the clusters have rates of these indexes that are significantly higher than the BUA average rate, and these are found consistently across all clusters in both 1971 and 2001. The only exception to this is socially rented housing in 1971. This was above average in the Catholic majority clusters of the Lower Falls and the Upper Falls and Ballymurphy as well as in the city center but was below the city average in other clusters. By 2001 these rates had changed, and rates of socially rented housing in all the clusters of violence were typically twice the BUA rate. By contrast, areas with very low rates of violence also appear to be very affluent using these indicators. In 1971 34 percent of households in areas that had a density of less than two killings per square kilometer had rates of households without a car of 34 percent and unemployment rates of less than 4 percent. In 2001 these rates had fallen to 21 percent for socially rented housing, 23 percent for lack of a car, and 4 percent for unemployment, all well below the city average.

There was, therefore, an interesting dynamic in patterns of violence in Belfast. There was a strong relationship between an area's levels of deprivation and its levels of violence. Beyond this, the nature of that violence, in terms of who perpetrated it and who the victim was, varied significantly but was closely related to the religious demography.

Interpreting Religious Change and Segregation

It is well established that the civil unrest of the Troubles led to large-scale population moves in Belfast.[4] These moves started in the late 1960s, a period before spatially detailed data are available and whose effects thus cannot be observed in this account; however, widespread housing moves as a result of violence and intimidation remained common in the early 1970s, a period that does fall within our analytical time period. The longer-term impact of these events will also have influenced residential location decisions later in the period through fear and ongoing but lower levels of violence and intimidation. Patterns of population change observed in Belfast are thus, to some extent, the consequence of sectarian factors and civil conflict as well as the more typical demographic processes described above.

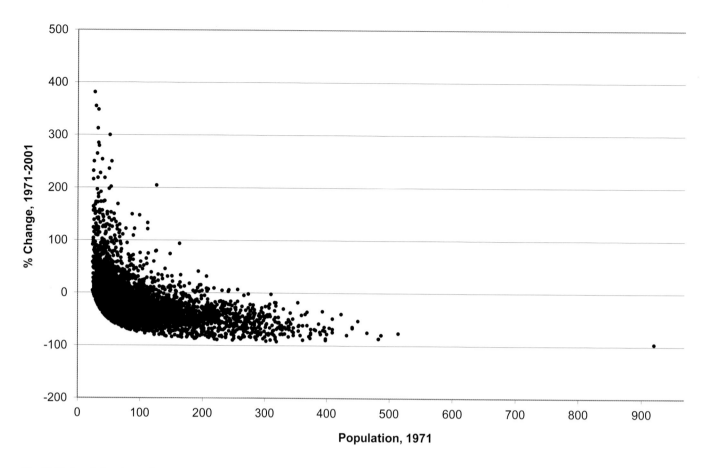

Fig. 11.10. Population change between 1971 and 2001 by population in 1971 for each 100 m grid cell. Note that cells with a population of fewer than twenty-five have been suppressed.

Figure 11.10 analyzes population change further and shows that, as with Northern Ireland as a whole, the areas that lost population between 1971 and 2001 tended to be the ones that were densely populated in 1971. The areas that were in the top 10 percent for population loss in the period from 1971 to 2001 had a mean population of over two hundred people per 100 m cell in 1971, whereas the areas that were in the top 10 percent for population growth in the same period had an average population of around seven people per 100 m cell in 1971. Likewise, the areas that tended to have higher rates of socially rented housing also tended to have high levels of non–car ownership in 1971. In 2001 the places that had gained the most population had higher employment levels, fewer people per household, and higher rates of owner-occupied housing. In short, the 1971 to 2001 period saw the demographic decline of areas that were inner city, densely populated, and relatively socially deprived and that experienced high levels of violence. The areas that saw the greatest growth were in 2001 better off on average, had higher rates of employment, and were less violent than places that had seen losses. With regard to housing tenure, there are interesting ambiguities, with areas with high owner occupation *and* relatively high levels of socially rented housing in 2001 in the top 10 percent of places for growth. This is a consequence of differential social and housing processes associated with demographic growth in Belfast's suburbs. The majority of this decentralization was associated with new private housing, but a significant proportion was also related to the creation of large public-housing estates on the edges of the city.

Were it not for violence, these patterns would be much as might be expected, given the experience of other cities. People tend to move away from

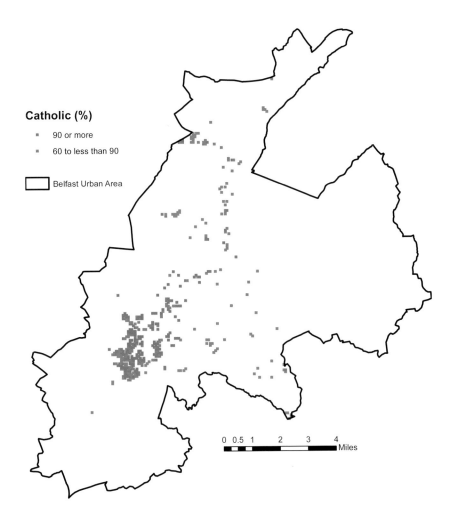

Fig. 11.11. Areas that were unpopulated in 1971 and more than 60 percent Catholic in 2001.

Catholic (%)

- 90 or more
- 60 to less than 90

☐ Belfast Urban Area

0 0.5 1 2 3 4
Miles

densely populated, deprived parts of cities to more affluent, less densely populated places.[5] Thus, population change in Belfast can be seen in the light of the broader European and North American experience. Moreover, Protestants and Catholics both seem to move in response to these stimuli. Both groups left areas that were densely populated in 1971, and the areas that saw the biggest population growth for both groups in 2001 tended to have populations that had higher employment and owner-occupation rates. In some places, both communities increased and decreased jointly. However, this kind of account ignores the generally different geographies of population change: Protestant increase seems to be located very much in the urban periphery, whereas Catholic decrease and increase are more spatially concentrated.

Figure 11.11 shows where the Catholic population has experienced the greatest proportionate growth and the context of this increase. The shaded cells were unpopulated in 1971 and were over 60 percent or 90 percent Catholic in 2001; in other words, these are areas where new housing, occupied by Catholics, had been built over the thirty-year period. Most of these areas are in the west of Belfast where there had been large public authority housing estates built in the 1970s and 1980s, although there is also some evidence of growth on the northwest and southeast axes of the city.

It is possible to say with some certainty that at least *some* of the changes seen in Belfast have been the direct result of fear, violence, and conflict, but it is equally possible to make a judgment that more normal forces, including planning policy, housing clearance in the inner city, and labor market change, have influenced the population of Belfast just as they have in many other cities. The problem is in deciding how much of the population changes are a direct result of sectarian fear and violence and how much were a result of more normal processes. This decision is made more problematic by the indirect impacts of fear both for individuals and for institutions. For individuals fear might be a conscious or an unconscious part of their decision making; for institutions such as government departments and planners there is a fine line between policies to promote mixing and those to promote separation, and some locational decisions (e.g., about planning the location of new housing) might have been prompted by the rationale that good fences make good neighbors. Beyond this, Catholics and Protestants as groups tended to be different in their socioeconomic characteristics—particularly so in the 1970s, although less so in 2001—and in 1971 they already tended to live separately in segregated places. The same social and economic forces acting upon the two groups may therefore have led to different geographical and demographic outcomes without recourse to the Troubles that Northern Ireland experienced in the 1970s, 1980s, and 1990s—unequal starting points will normally create unequal results in the absence of strong attempts to produce, in this case, population mixing.

Conclusions

Belfast changed markedly between 1971 and 2001, becoming more Catholic, more segregated, older, and less densely populated. In some ways it was more affluent: owner occupation rose, and more households gained access to private transport. In other ways the economic position of some residents worsened as employment rates fell, particularly in the inner city. Belfast might, therefore, be thought to have become more socioeconomically polarized—an unsurprising finding, given the evidence for growing inequalities both in the U.K. and in Ireland.[6] The processes that had brought the city to this point were, to some degree, like those experienced elsewhere in Europe and North America. Belfast, for example, was hardly the only city that had seen the devastation of its manufacturing base. The most dramatic changes were seen between the censuses of 1971 and 1991, and—although it is impossible to date them with precision—we know that the 1970s saw large-scale economic and demographic changes across the developed world. It is tempting to assume the same timing in the case of Belfast. It is difficult to be certain about the forces that have led to changes in Belfast's population geographies. Without doubt, sectarian factors, including the violence and fear associated with the Troubles, were important; however, placing these forces in the wider context of urban change indicates that other influences also worked upon Belfast, and it is important to understand these too.

Conclusions: Ireland's Religious Geographies—Stability or Change? 12

Looking Backward

Figure 12.1 shows the distribution of Catholics in 1834 and compares this with their distribution in 2001/2002 as interpolated onto 1834 Church of Ireland dioceses. In many ways very little has changed: Catholics make up the vast majority of the population over much of the island with the exception of Ulster, especially east Ulster, where they are often a minority, Dublin and the Pale, and parts of the south, especially around Cork. In these areas there are significant Protestant populations. These two maps actually understate the degree of long-term spatioreligious stability—while 1834 gives us the first detailed head counts, these geographies were actually laid down in the plantation period over two centuries earlier. This period of emigration and colonization from Britain into Ulster, Dublin and the Pale, and Munster left the geographies that are still clearly apparent today. It also left a legacy of economic and social division and interdependence between the two groups and an intertwining of religion, ethnonational identity, and political opinion that has periodically flared into violence. Conversely, these maps may also overstate the degree of stability, because dioceses are very aggregate units. The maps can and do conceal patterns of concentration and more localized change that more detailed geographies, were they available, might reveal. Nevertheless, this degree of stability appears remarkable. From 1834 to 2001/2002 only three dioceses containing approximately 7.5 percent of the population at both dates saw the Catholic proportion of their population change by more than ten percentage points. Of these, Raphoe and Derry in west Ulster saw their Catholic populations rise from 70 percent to 86 percent and 54 percent to 65 percent, respectively, while the tiny diocese of Kilfenora on the southern shore of the west coast's Galway Bay saw its Catholic population decline from 99 percent to 84 percent.

The nineteenth and twentieth centuries saw huge social and economic changes. Some have been long-term, gradual, and mainly associated with broad demographic and economic trends. Others have been sudden, dramatic, and frequently highly traumatic, including the Great Famine, Partition and the Civil War, and the Troubles. All these changes have had profound demographic consequences for the island's population. In many cases religious demography and its links to identity, politics, and socioeconomic status have also, at least in the past, been causes of the changes. Despite all of these developments, spatioreligious patterns appear to have remained highly constant.

Fig. 12.1. Catholic populations in (*a*) 1834 and (*b*) 2001/2002. The 2001/2002 data have been interpolated onto the dioceses used by the 1834 Royal Commission.

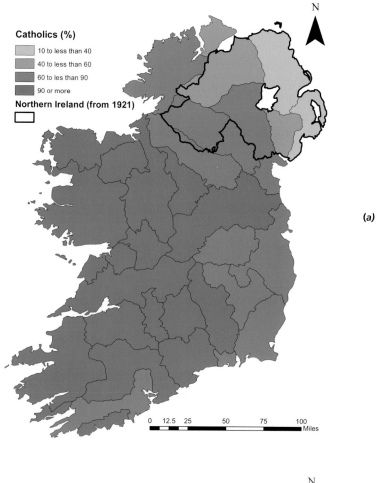

(*a*)

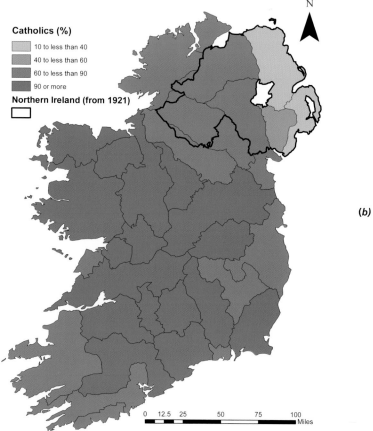

(*b*)

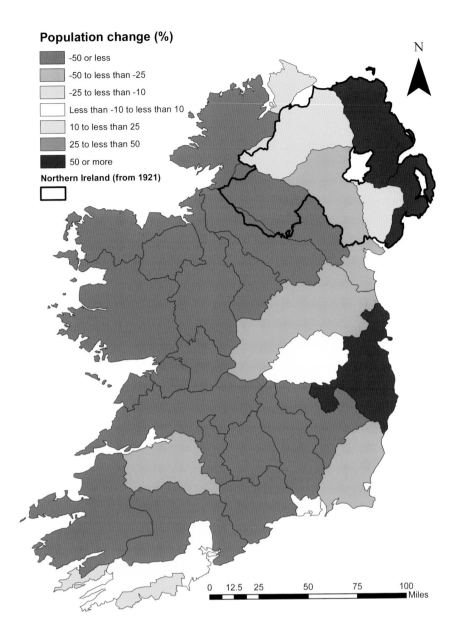

Population change (%)

- -50 or less
- -50 to less than -25
- -25 to less than -10
- Less than -10 to less than 10
- 10 to less than 25
- 25 to less than 50
- 50 or more

Northern Ireland (from 1921)

N

0 12.5 25   50   75   100
Miles

Fig. 12.2. Percentage of population change between 1834 and 2001/2002. The 2001/2002 data have been interpolated onto the dioceses used by the 1834 Royal Commission.

While there has been much stability, it is wrong to say that there has been little change. Long-term demographic change has had a marked impact on the geography of Ireland. Overall, the population fell by 28 percent from 1834 to 2001/2002. This change was driven particularly by emigration rates over the long term, although the catastrophic events of the Famine years also made a major contribution. The effects of this loss have not been even, as figure 12.2 shows. Three dioceses—Connor, Down, and Dublin— saw large population increases of 73 percent, 88 percent, and 156 percent, respectively, as a result of the growth of Belfast and Dublin. Of the remaining twenty-nine dioceses, the populations of three changed by less than 10 percent, while twenty—covering the vast majority of the west and the midlands—have shown population declines of over 50 percent. While population change can be difficult to interpret, as it combines the effects of out-migration, in-migration, births, and deaths, given the relative lack of immigration over most of the nineteenth and twentieth centuries, it seems likely that a major cause of this pattern has been people migrating out of the rural areas that were once and still remain overwhelmingly Catholic.

Some of this out-migration has been emigration, which has contributed to the overall Catholic proportion of the population declining from 80.7 percent to 74.7 percent. Some of the rest has been rural to urban migration into the cities and towns that were the Protestant heartlands in the early nineteenth century. This migration has contributed to these areas having increasingly large Catholic populations. Connor, which approximates to County Antrim, saw its Catholic share of the population rise from 26.4 percent to 35.8 percent, while in County Dublin the Catholic share rose from 77.9 percent to 84.2 percent. This means that in Belfast and Dublin, in particular, demographic change over the two centuries has meant that Catholics and Protestants now live in much closer proximity than they did previously.

Living closer together does not, however, mean that the two communities have necessarily become closer. While demographic change shows a clear divide between the rural west and the more urban east, from the perspective of religious identity there has always been a clear divide between north and south. As we have seen, religious identity, ethnonational identity, and political agendas came together in Ireland in the early twentieth century, leading to the predominantly Catholic, Gaelic-Irish, proindependence south becoming separated from the predominantly Protestant, British-Irish, prounion north. Both religion and society in the two parts of the island have subsequently followed very different trajectories. The south saw economic stagnation for much of the twentieth century followed by rapid economic growth in the last few decades that had its origins in the reforms of the 1960s. The north saw its economic primacy based on manufacturing disappear as these traditional industries declined internationally.

Figure 12.3 shows the demographic trends of the minority religions in the two parts of Ireland. The Protestant minority in the south contributed a little over 10 percent of the population until Partition. Immediately after Partition there was a sharp decline in the Protestant minority, which continued, albeit more slowly, until 1961, when the non-Catholic share of the Republic's population stood at only 5.1 percent. Since then the non-Catholic share of the population has increased significantly and in 2002 had risen to pre-Partition levels. This increase has not been driven by an increase in the size of the traditional Irish Protestant community; instead, it reflects the secularization and immigration that have accompanied economic progress. In Northern Ireland, conversely, the Catholic population declined in the pre-Partition period but stabilized afterward at around 34 percent. While there was a slight upward trend in this population until 1961, the start of the Troubles seems to have coincided with, and probably led to, a sharp loss of Catholics from the north. As discussed in chapter 9, figures for Catholics from the 1981 Northern Ireland census are unreliable, but by 1991 an upward trend had resumed, and by 2001 the proportion of Catholics in these six counties was higher than it had been since the nineteenth century.

Taken at face value, when we consider the prevailing social climates in which the two minorities existed, these trends are perhaps surprising. The Protestant minority in the south, while tiny in comparison to Catholics north of the border, has been generally well tolerated since the hostilities of

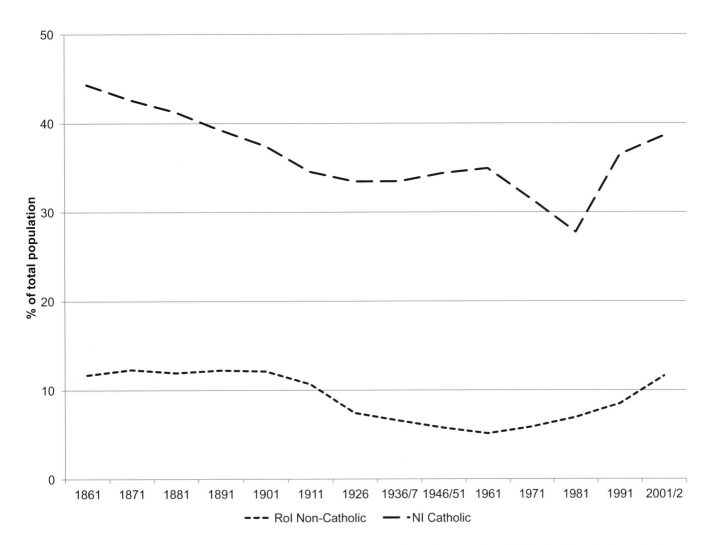

--- Rol Non-Catholic — ·NI Catholic

the Partition era subsided, yet the number of Protestants declined through-out much of the period. Life for many Catholics in Northern Ireland, by contrast, has for much of the time been characterized by discrimination and hostility, despite which their numbers have grown. It might be specu-lated that this increase demonstrates that the economic forces keeping Catholics in Northern Ireland were more powerful than the disadvantages of life in the region, apart from a short period at the start of the Troubles.[1] But this speculation fails to provide an explanation for the entire trend, as much of the increase in the Catholic population of Northern Ireland took place in the period after 1971, when the province was in economic decline and lower-level antagonism had given way to the outright violence of the Troubles.

So why has there been major political and sectarian conflict in the north while the south has been peaceful? Part of the explanation may lie in the changing roles of religion both north and south. In the Republic the importance of the link between religion, national identity, and poli-tics seems to have largely disappeared. With no prospect of the Free State rejoining the U.K., the Anglo-Irish identity of many southern Protestants declined in importance, particularly as many of them felt betrayed by their northern brethren.[2] At the same time the political agenda moved away from the national question to focus instead on internal concerns within

Fig. 12.3. The size of the minority populations of the two parts of Ireland from 1861 to 2001/2002. RoI non-Catholic: non-Catholics in the (modern) Republic of Ireland; NI Catholic: Catholics in (modern) Northern Ireland. Note that the 1981 figure for Catholics in Northern Ireland must be treated with caution due to problems with this census.

the Free State and the Republic.[3] In the north, by contrast, differences of politics, culture, and identity still break down largely along religious lines, and the consequences of this breakdown have been tragic.

Future Trajectories: Moving Apart or Moving Together?

This book finishes with the censuses of 2001/2002, a time when both parts of Ireland were very much at a crossroads. In Northern Ireland the peace process was still in its early stages. The signs were encouraging, but there were still major questions about whether the process would lead to a lasting peace. The Republic, by contrast, showed a very optimistic picture, with the Celtic Tiger economy having achieved what appeared to be little short of an economic miracle. A decade later the situation is, if anything, more confused. The peace process in Northern Ireland has, despite some setbacks, been solid. The political power-sharing system that it brought in also appears relatively stable; indeed, at one stage during the 2000s Ian Paisley, the hard-line leader of the Democratic Unionist Party, and Martin McGuiness, a former senior IRA leader, were, respectively, the leader and deputy leader of the Northern Ireland Assembly. The working relationship of these former enemies appeared so close that they were nicknamed the "Chuckle Brothers."[4]

At a more day-to-day level, however, the situation is less encouraging. While major efforts have been made to outlaw discrimination in areas such as employment, policing, and housing, low-level sectarianism remains a significant problem. Peace lines still scar the urban landscape, sectarian intimidation and violence are still facts of life, and rioting still commonly accompanies the Orange Order marching season in July. In the summer of 2011 Northern Ireland witnessed the worst civil disorder since the "end" of the Troubles.[5] Beyond this, many public services, such as the provision of schools and leisure facilities, have to explicitly acknowledge the sectarian divide by offering duplicate facilities in Catholic and Protestant areas. In short, spatioreligious divides are still in many ways as stark as they were during the Troubles.

At the same time, it is beginning to become true to say that there is no longer a clear zero-sum game in Northern Ireland's religions between Catholics and Protestants. Recent years have seen immigration into Northern Ireland, albeit at a far lower level than other parts of the U.K., and there are increasing minorities from East Asia, South Asia, and eastern Europe in particular. Accurate figures for the numbers of these people and their locations are unavailable, as, at the time of writing, the results of the 2011 census had yet to be published. The transition toward a degree of cosmopolitanism has not, however, been painless, and there have been a number of ugly incidents aimed at immigrants and minority groups.[6] Some commentators have attributed these incidents to the legacy of the Troubles, which consolidated a "culture of intolerance" within some sections of Northern Irish society.[7] Nevertheless, immigration, the rise in secularization, and ongoing political developments that reduce the lure of hard-line Irish nationalist and unionist politics may help to break down geographical and identity barriers. Whatever happens in the longer term, Northern Ireland will not return to the failed attempt to be a Protestant state for a Protestant people in which Protestantism not only was seen as a religion but was synonymous with being pro-British and prounion. Instead, Northern Ireland will have to adapt to having two major cultures and a significant number of more minor ones. If religion on both sides of the divide continues to dominate community identity and political opinion, this adaptation may prove difficult.

If all groups are able to benefit, then economic success has the potential to be an important factor in turning Northern Ireland into a more cohesive society. There have been encouraging signs for its economy, as Belfast, in particular, has emerged from the Troubles and reinvented itself as a postindustrial city in a way similar to that of places such as Glasgow, Manchester, and Liverpool. Regeneration within the city center and the transformation of derelict shipyards into the Titanic Quarter are testament to this.[8] However, Northern Ireland's economy still remains heavily dependent on the public sector, and the contraction of public-sector spending after the financial crisis of 2008 and the subsequent election of a U.K. government that sees deficit reduction as a key priority have had major implications for Northern Ireland's prosperity. Whether a city that has been heavily dependent on government spending for several decades can develop itself based on a reinvented and thriving private sector remains to be seen. A failure to do so may, however, have significant implications not just for the economy but also for a cohesive society.

If the financial crisis has presented long-term challenges for Northern Ireland, its impacts in the Republic were immediate and severe. In the lead-up to 2008 the Celtic Tiger seemed unstoppable. It brought with it a successful, rich, and diverse economy that turned an ethos of emigration into one of immigration. As with Northern Ireland, it is currently difficult to be precise about the impact of immigration on the Republic's religious geographies, but even a casual visitor to Dublin cannot fail to notice how multicultural the city is. At the same time, the child abuse scandals within the Catholic Church and the ongoing rise in secularization have continued to weaken the influence of Catholicism. Thus, the days when the Republic was an overwhelmingly Catholic state, both in terms of numbers and, more particularly, in the power of the church, have gone and are unlikely to be returned to. The link between Catholicism and nationality, which was forged in the nineteenth century, has effectively been broken. Similarly, the sense of Irish identity, which largely conceived of itself in a post-colonial opposition to Britishness, has largely disappeared as the Irish state and its economy have matured. With this, the geographies of religion have become increasingly irrelevant. Thus, in the Republic and in Northern Ireland, the old certainties provided by religion have been deconstructed, to be replaced by a more inclusive but also more hesitant form of identity.

The Celtic Tiger economy brought with it a housing boom. When the economy collapsed in the financial crisis, Ireland's banks needed to be rescued by the government. The bond markets deemed this rescue to be unsustainable, so the government, in turn, needed to be bailed out by the European Union (EU), leaving Ireland as one of the most indebted economies in Europe. The long-term effects of the crisis are, of course, impossible to predict at present, but it has already ushered in an unwelcome return to emigration. However, while still heavily dependent on FDI, the Irish economy is perhaps in a better position than comparably indebted states in that it has attracted not only assembly operations to the country but also a significant value-added research and development infrastructure.

So far this discussion has concentrated on two Irelands rather than one. These two have had different trajectories during the twentieth century, and we have speculated on their individual futures. This speculation poses the question of whether there is any chance of a single Ireland reemerging. At a political level this seems unlikely. As part of the peace process the U.K. and Republic of Ireland governments and the major paramilitary organizations all recognized that Northern Ireland's electorate should have the right to determine the province's future as a state. This makes it a near certainty that Northern Ireland will continue to be part of the U.K. for the foreseeable future. We may, however, be moving toward a situation where the island follows a more unified trajectory in the twenty-first century than it has in the twentieth. Formal cross-border links are increasing, and informal evidence suggests that more and more people are crossing the border for business and for pleasure. Links between the U.K. and the Republic have become ever stronger, as symbolized by the success of Queen Elizabeth's visit in 2011, the first state visit by a British monarch since George V in 1911.[9] Increasingly as well, Ireland and the U.K. have become partners within the EU, developing ever stronger economic and political ties and thus helping to close the constitutional divide between nationalism and unionism. Economically, the old dichotomy between an agricultural south and a manufacturing-based north has collapsed as the economies of both states have increasingly converged around the service sector, increasing a degree of interdependence between the two that is only likely to grow further. Thus, while the border will remain, the rift that it represented may be diminishing as the religious, identity, political, and economic divides that created it become less important.

Notes on Methods and Literature: From Historical GIS Databases to Narrative Histories

This book sits astride two approaches to history. It presents a traditional argument in response to perennial questions in Irish history, but it does so using a nontraditional method. By tapping the power of new geospatial technologies, the book is able to offer a different perspective on the intersection of religion, politics, and power in Irish history—a perspective that stresses the importance of space. What follows is not a standard bibliography or a discourse on method, both of which could be the subject of separate books.[1] Rather, it is a brief reflection on the issues raised by this approach and a suggestion about where it fits within the scholarly literature.

Space, Time, and Theme in Irish Histories

A monograph covers a range of different but related topics that are brought together to create and structure a coherent overarching story. To achieve this, histories frequently use either time, theme, or a combination of both to structure their central narrative. Perhaps the most common approach is for a book to tell the story of change over a period of time. To achieve this approach the book is subdivided into chapters that cover shorter time periods within which authors discuss multiple themes. Examples of this approach from Irish histories include R. F. Foster's *Modern Ireland 1600–1972*; C. J. Beckett's *The Making of Modern Ireland, 1603–1923*; F. S. L. Lyons's *Ireland since the Famine*, which splits Ireland spatially into north and south after Partition; and, for a shorter time period, M. Laffan's *The Partition of Ireland 1911–1925*.[2] Theme can also be used to derive the main structure. J. Mokyr's *Why Ireland Starved: A Quantitative and Analytical History of the Irish Economy, 1800–1850* uses this approach.[3] It includes chapters on themes such as population; land, leases, and tenures; and the economics of rural conflict. Other histories, including M. E. Daly's *The Slow Failure: Population Decline in Independent Ireland, 1920–1973* and C. Ó Gráda's *Ireland: A New Economic History*, use a combination of both time and theme.[4] Daly's chapters cover themes such as rural Ireland, the Irish family, and emigration, each of which is explored over three major time periods, while Ó Gráda's book includes chapters titled "Industry 1870–1914: An Overview," "Industry 1870–1914: Problems and Prospects," and "Banking and Industrial Finance: 1850–1939."

In this way the stress on time and theme—in different proportions—allows an author to present a coherent narrative. The problem is, of course, the neglect of space. While most of the titles listed above contain the word "Ireland" or "Irish," geography is basically only used as a container and is rarely raised in any other context apart from in a very minor way, such as occurs in Lyons to tell the differences between Northern Ireland and the Free State/Republic after Partition.[5] The difficulty is that this implies that there is only one story (diverging into two in Lyons's case) and that this story occurred in much the same way in all places.[6] In actual fact, this single story results from the aggregation of the combined experiences of many local communities. As E. A. Wrigley and R. S. Schofield's *Population History of England 1541–1871: A Reconstruction* notes, the aggregate experience can be thought of as an average that may only reflect the actual experience of a very few individual communities.[7]

An alternative approach to structuring a history that is more space-centric is to use an atlas. In the early days of Historical Geographical Information Systems (HGIS), use of an atlas was seen as an obvious way of turning the database into a book. Irish examples include L. Kennedy et al.'s *Mapping the Great Irish Famine: An Atlas of the Famine Years* and J. Gleeson et al.'s *The Atlas of the Island of Ireland: Mapping Social and Economic Change*, which, despite its title, only covers the period from 1991 to 2001/2002.[8] One issue atlases of this type have is that they are limited in their ability to manage time because they are based around choropleth maps. A second, more serious issue is that they fail to provide an overarching narrative, instead resorting to short sections of text to accompany sets of maps covering a wide range of different themes and, perhaps, dates. Thus in their attempt to describe space, atlases tend to sacrifice narrative and as a consequence often become largely descriptive. In their *England on the Eve of the Black Death: An Atlas of Lay Lordship, Land and Wealth, 1300–49*, another HGIS-based atlas, B. M. S. Campbell and K. Bartley are commendably honest about this limitation, stating: "No effort is made to interpret the maps and graphs. To have done so would have required another entire volume and a great deal of additional research. Rather, they are left to speak for themselves. Hopefully other historians will use and interpret them in whatever ways they find most useful."[9]

Space, Time, and Theme in Geographical Information Systems

The field of HGIS originated in the late 1990s.[10] Its early emphasis was largely on technology, but it has increasingly evolved to become more focused on developing applied scholarship in history and in the humanities more broadly, leading to a shift toward the closely related fields of spatial history, spatial humanities, and geohumanities.[11] While HGIS and spatial history have delivered a wide range of high-quality journal articles and book chapters, to date the field has been less effective at delivering monographs.[12] There are a few notable exceptions to this, including B. M. S. Campbell's *English Seigniorial Agriculture, 1250–1450*, B. Donahue's *The Great Meadow: Farmers and the Land in Colonial Concord*, G. Cunfer's *On the Great Plains: Agriculture and Environment*, and C. Gordon's *Mapping Decline: St. Louis and the Fate of the American City*; however, as the monograph still represents the "gold standard" in many humanities disciplines, the need to take the knowledge contained in HGIS databases and turn it into monographs still represents a major challenge to the field.[13]

It is noticeable that three of the monographs listed above are concerned with agriculture and land use, while the fourth is an urban history.

Equally noticeable is that none have originated from National Historical Geographical Information Systems (NHGIS) projects. The construction of these databases was one of the early trends within HGIS. They are large, complex, and costly systems that usually hold a country's census and related data over the past two centuries or so. Good examples include the Great Britain Historical GIS, the U.S. National Historical GIS, and the Historical GIS of Belgian Territorial Structure.[14] The database on which this book is based contains Irish census data from the first census in 1821 to the most recent in 2001/2002 and is typical of these databases. The data in an NHGIS, as with many other HGIS databases, have three components, all of which bring their own complexities and all of which need to be used effectively. First they have a thematic component—termed *attribute data* in GIS jargon—which provides information on subjects such as population, religion, industry, education levels, and so on; these data are usually represented quantitatively. Second, they have a spatial component, which, for most of the data in an NHGIS, is the administrative units—represented by *polygons*—for which the data apply. Finally, the third component is time, which for census data is in the form of snapshots of society taken around every decade or so. Thus, in theory, an NHGIS allows us to take multiple, complex, interrelated attributes and explore how they vary over space and time. Going one stage further, this process in turn should allow narratives to be produced that explore theme, time, and space, thus addressing the lacuna identified in the previous section: the lack of a spatial component within most narrative histories.

In reality, doing this is not as simple as it might appear. Even in the early days of the development of GIS it was understood that GIS databases were designed to store, analyze, and visualize the spatial and attribute components of data, but they were not designed to handle time. As far back as the late 1980s considerable research effort was being put into adding time to GIS.[15] In an early piece of work on the subject, G. Langran and N. R. Chrisman argued that traditionally to represent one of these three components, the second is controlled, and the third is fixed.[16] They give census data as one example: time is fixed by conducting the census on a single night, space is controlled by dividing the country into arbitrary administrative units, and this enables theme—the number of people living in the unit on that night—to be measured accurately. Weather reports use a different structure. They fix location by recording readings at fixed sites, time is controlled as readings are taken over certain intervals such as daily, and theme—the maximum and minimum temperatures or amount of rainfall—is measured. Soils maps provide another alternative. Here time is fixed, theme is controlled by dividing the soils into types, and space is measured to identify where the boundaries between types appear. Langran and Chrisman's argument is that GIS should make this approach unnecessary and that it should be possible to conduct analyses through all three components simultaneously.

More recent literature makes it clear that progress toward this ideal has been limited.[17] This is probably because although it is technically possible for a computer to perform an analysis of theme, space, and time simultaneously, human perception actually requires us to simplify data and findings to make them understandable. This simplification results in us tending to take either space-centric or time-centric approaches. A simple example is provided by the problems in visualizing change over space and time in a comprehensible way. Choropleth maps, such as frequently occur in this book, are space-centric in that they stress space but fix time either to a single snapshot or to the change between two, and only two, dates. Additionally, to aid perception, these maps also have to simplify attribute by dividing the statistical data into a small number of arbitrary classes, each of which is given its own distinct shading.[18] Thus, time is fixed and attribute is simplified to allow space to be stressed. The time-centric alternative is to use time-series graphs, which allow us to explore how attribute values change over time, thus emphasizing both time and attribute but doing so at the expense of space. A small number of time series can be used to compare different places (e.g., all of Ireland, Northern Ireland, and the Republic); however, their ability to properly subdivide space is almost nonexistent.

There are few, if any, effective compromises between these two approaches. Superficially, animating choropleth maps appear to combine both; however, as we have already considered, for choropleth maps to be perceived effectively, they need to simplify attribute. Adding the extra complexity of time means that further simplification of space, time, attribute, or combinations of them will be required. For complex situations such as the changing distribution of a religion from 1861 to 2001/2002, simplifying the patterns to make them understandable while preserving sufficient complexity to make the patterns meaningful is unlikely to be possible. A further, practical point is that animations cannot be published on paper, and paper remains the most popular format for publishing and reading monographs. A simpler approach might be to put a time-series graph in each polygon on a choropleth-style map. As with animations, this is technically straightforward; however, from the perspective of human perception, it is unlikely to be satisfactory for more than a few polygons. Therefore, while there is the technical ability to produce forms of visualization that include space, time, and theme simultaneously, human perception still means that we are largely confined to space-centric or time-centric approaches.

The division between time-centric and space-centric approaches also occurs at the academic level. Despite the fact that there have been many papers from geographers, in particular, calling for space and time to be investigated together,[19] in broad disciplinary terms history is time-centric and tends to stress periods in the past and stories of temporal change, while geography is space-centric and focuses on places and the relationships between them. Historical geography attempts to cross this divide but often follows an approach that focuses on the geographies of a period in the past, or it looks at how geographies—especially landscapes—have changed over time, approaches that stress space and time, respectively.[20] Thus, the situation remains that at both detailed technical and broad academic levels, drawing together space, time, and theme remains a problem.

Space, Time, and Theme in This Book

This book was based primarily on an HGIS database of Irish historical statistics—mainly census data—covering the period from 1821 to 2001/2002.

Its narrative is an attempt to combine space, time, and theme by combining the spatial complexity presented in an atlas with a coherent temporal narrative that explains how different themes related to each other and how and why they changed over time. As is typical of a historical monograph, it is structured primarily by time, with the chapters addressing what we consider to be the major periods in nineteenth- and twentieth-century Irish history and an earlier chapter providing the historical background. The exception comes at the end of the book, by which point we consider that Northern Ireland and the Republic have diverged sufficiently to need to be treated separately. Beneath the overarching temporal structure, the chapters subdivide into major themes such as population, religion, industry, emigration, and so on. Within this structure, the maps provide the spatial information that allows us to explicitly address geographical variation within the narrative.

Our emphasis on time contrasts with most other monographs that have emerged from HGIS, most of which largely follow a theme-based structure: Campbell and Cunfer organized their books on themes such as arable land, grazing land, crop diversity, and so on.[21] Gordon is similar, concentrating on themes such as race, zoning, and urban renewal, while some of Donahue's themes also have a temporal aspect, concentrating on, for example, different periods of land use.[22] Time is not central to any of them, despite the fact that they cover periods varying from sixty years to two centuries. Space is important to all of them, but, as with this book, space is usually exploited by using maps as the device from which to explore geographical variation.

Our data contain around three hundred spatial units for eighteen snapshots through time with multiple and varying attributes that are sometimes only available at county level. Moving from these data into knowledge and narrative that fully exploits theme, time, and space still represents a challenge and, indeed, may not be fully achievable or even desirable. At a technical level, figures still need to simplify and even fix one or more of these attributes to make variations in the others more understandable. At the academic level, producing a coherent narrative requires a structure that privileges one of these attributes, often either theme or—as in this book—time. The trick, it seems to us, is that this overarching structure, in which the main message of the narrative is revealed, should enable the complexity and diversity within all the others to emerge within their broader context rather than to ignore, oversimplify, or deaden one or more of the three. Thus, while it is not possible to move to an approach that fully integrates theme, time, and space, all three can still be included to greater or lesser extents. Importantly, simply ignoring one of the three—typically space by historians or time by geographers—is both undesirable and unnecessary.

Notes

1. Geography, Religion, and Society in Ireland

1. Conflict Archive on the Internet (CAIN) has many good photographs of examples of this (http://cain.ulst.ac.uk/photographs/index.html, accessed 7 February 2012). See, for example, http://cain.ulst.ac.uk/images/photos/coantrim /ballymena/BMA05CHUharryville2.htm and http://cain.ulst.ac.uk/images/photos/belfast /murloy10.htm#murloy10 (accessed 7 February 2012) for photographs from Protestant areas and http://cain.ulst.ac.uk/images/photos/belfast/falls /fallsmura11.htm#fallsmura11 and http://cain.ulst .ac.uk/images/photos/derry/bogsideartists /muralbattle2.htm#muralbattle2 (accessed 7 February 2012) for photographs from Catholic areas.

2. A. Walsham, *The Reformation of the Landscape: Religion, Identity, and Memory in Early Modern Britain and Ireland* (Oxford: Oxford University Press, 2011), 544–47.

3. A. D. Smith, *Chosen Peoples: Sacred Sources of National Identity* (Oxford: Oxford University Press, 2003), 154; and D. H. Akenson, *God's Peoples: Covenant and Land in South Africa, Israel and Ulster* (Ithaca, N.Y.: Cornell University Press, 1992), 183–202.

4. Again, CAIN provides good images at http:// cain.ulst.ac.uk/photographs/index.html. See, for example, http://cain.ulst.ac.uk/images/photos /belfast/peaceline/peaceline1.htm#peaceline1 and http://cain.ulst.ac.uk/images/photos/belfast /peaceline/lanark1.htm#lanark1 (accessed 7 February 2012).

5. D. H. Akenson, *Small Differences: Irish Catholics and Irish Protestants 1815–1922* (Montreal: McGill-Queen's University Press; Dublin: Gill & Macmillan, 1991), 16.

6. D. Ferriter, *The Transformation of Ireland 1900–2000* (London: Profile Books, 2004), 662–63.

7. These are available as L. A. Clarkson, L. Kennedy, E. M. Crawford, and M. E. Dowling, *Database of Irish Historical Statistics 1861–1911* (computer file) (Colchester, Essex: UK Data Archive [distributor], November 1997, study number 3579); and M. W. Dowling, L. A. Clarkson, L. Kennedy, and E. M. Crawford, *Database of Irish Historical Statistics: Census Material, 1901–1971* (computer file) (Colchester, Essex: UK Data Archive [distributor], May 1998, study number 3542).

8. I. N. Gregory and P. S. Ell, *Historical GIS: Techniques, Methodologies and Scholarship* (Cambridge: Cambridge University Press, 2007), 9–12.

9. I. N. Gregory and P. S. Ell, "Breaking the Boundaries: Integrating 200 Years of the Census Using GIS," *Journal of the Royal Statistical Society Series A* 168 (2005): 419–37.

10. See A. S. Fotheringham, C. Brunsdon, and M. Charlton, *Quantitative Geography: Perspectives on Spatial Data Analysis* (London: Sage, 2000) for a general overview. I. N. Gregory, "'A Map Is Just a Bad Graph': Why Spatial Statistics Are Important in Historical GIS," in *Placing History: How Maps, Spatial Data and GIS Are Changing Historical Scholarship,* ed. A. K. Knowles (Redlands, Calif.: ESRI Press, 2008), 123–49, describes how these approaches have been used in historical studies.

11. R. White, *What Is Spatial History?,* http:// www.stanford.edu/group/spatialhistory/cgi-bin /site/pub.php?id=29 (accessed 25 March 2010).

12. A. R. H. Baker, *Geography and History: Bridging the Divide* (Cambridge: Cambridge University Press, 2003).

13. See D. Massey, "Space-Time, 'Science' and the Relationship between Physical Geography and Human Geography," *Transactions of the Institute of British Geographers,* n.s., 24 (1999): 261–76; and Massey, *For Space* (Sage: London, 2005), for the importance of handling space and time together. See Gregory and Ell, *Historical GIS,* chap. 6, for a discussion of how GIS can help to apply this in historical research.

14. Gregory and Ell, *Historical GIS,* 100–105.

15. M. Sutton, *An Index of Deaths from the Conflict in Ireland,* http://cain.ulst.ac.uk/sutton /book/ (accessed 25 March 2010).

2. The Plantations

1. R. F. Foster, *Modern Ireland 1600–1972* (Harmondsworth: Penguin, 1989), 59.

2. G. A. Hayes-McCoy, "The Royal Supremacy and Ecclesiastical Revolution, 1534–47," in *A New History of Ireland III: Early Modern Ireland 1534–1691,* ed. T. W. Moody, F. X. Martin, and F. J. Byrne (Oxford: Clarendon Press, 1976), 39.

3. A. Cosgrove, "The Gaelic Resurgence and the Geraldine Supremacy, c. 1400–1534," in *The Course of Irish History,* ed. T. W. Moody and F. X. Martin, 4th ed. (Cork: Mercier Press, 2001), 133.

4. D. B. Quinn and K. W. Nicholls, "Ireland in 1534," in Moody, Martin, and Byrne, *A New History of Ireland III,* 1.

5. G. A. Hayes-McCoy, "The Tudor Conquest 1534–1603," in Moody and Martin, *The Course of Irish History,* 140.

6. A. Clarke, "The Colonisation of Ulster and the Rebellion of 1641," in Moody and Martin, *The Course of Irish History,* 153.

7. G. A. Hayes-McCoy, "Conciliation, Coercion, and the Protestant Reformation, 1547–71," in Moody, Martin, and Byrne, *A New History of Ireland III,* 79.

8. For an account of Edward VI's policies, see ibid., 69–70.

9. Ibid., 69.

10. Ibid., 79.

11. G. A. Hayes-McCoy, "The Completion of the Tudor Conquest, and the Advance of the Counter-Reformation, 1571–1603," in Moody, Martin, and Byrne, *A New History of Ireland III,* 108–109.

12. Ibid., 114.

13. R. Lacey, *Robert, Earl of Essex: An Elizabethan Icarus* (London: History Book Club, 1970), 216.

14. Foster, *Modern Ireland,* 67, 70.

15. M. MacCarthy-Morrogh, *The Munster Plantation: English Migration to Southern Ireland, 1583–1641* (Oxford: Clarendon Press, 1986), 276.

16. N. P. Canny, *The Upstart Earl: A Study of the Social and Mental World of Richard Boyle, First Earl of Cork, 1566–1643* (Cambridge: Cambridge University Press, 1982), 134.

17. Foster, *Modern Ireland,* 7.

18. A. Clarke, "Pacification, Plantation, and the Catholic Question, 1603–23," in Moody, Martin, and Byrne, *A New History of Ireland III,* 200.

19. T. W. Moody, *The Londonderry Plantation* (Belfast: William Mullan & Son, 1939), 35.

20. J. C. Beckett, *The Making of Modern Ireland 1603–1923* (London: Faber & Faber, 1966), 47.

21. See T. W. Moody, F. X. Martin, and F. J. Byrne, eds., *A New History of Ireland Volume IX: Maps, Genealogies, Lists* (Oxford: Clarendon Press, 1984), 44, 46, 50–51; and Moody, *Londonderry Plantation.*

22. Foster, *Modern Ireland,* 5.

23. Ibid., 63.

24. Clarke, "Pacification," 202.

25. Foster, *Modern Ireland,* 63.

26. H. O'Sullivan, "Dynamics of Regional Development: Processes of Assimilation and Division in the Marchland of South-East Ulster in Late Medieval and Early-Modern Ireland," in *British Interventions in Early-Modern Ireland,* ed. C. Brady and J. Ohlmeyer (Cambridge: Cambridge University Press, 2005), 66.

27. Foster, *Modern Ireland,* 72; R. Gillespie, "The Native Irish and the Plantation of Ulster," in "Across the Narrow Sea: Plantations in Ulster," unpublished conference proceedings, Four Seasons Hotel, Monaghan, 7 and 8 November 2008.

28. P. J. Duffy, "Social and Spatial Order in the MacMahon Lordship of Airghialla in the Late Sixteenth Century," in *Gaelic Ireland: Land, Lordship and Settlement c. 1250–c. 1650,* ed. P. J. Duffy, D. Edwards, and E. FitzPatrick (Dublin: Four Courts Press, 2001), 122.

29. A. Clarke, "The Irish Economy, 1600–60," in Moody, Martin, and Byrne, *A New History of Ireland III,* 176.

30. Moody, *Londonderry Plantation,* 185.

31. Clarke, "Pacification," 203.

32. K. M. Brown, "Courtiers and Cavaliers: Service, Anglicisation and Loyalty among the Royalist Nobility," in *The Scottish National Covenant in Its British Context 1638–51,* ed. J. Morrill (Edinburgh: Edinburgh University Press, 1999), 185.

33. J. Cunningham, "From the Broads to the Lakelands: English Settlers in Fermanagh," in "Across the Narrow Sea."

34. Clarke, "Pacification," 204.

35. M. Elliott, *The Catholics of Ulster* (London: Penguin, 2001), 195.

36. T. W. Moody, "Irish History and Irish Mythology," *Hermathena* 124 (1978): 10.

37. Foster, *Modern Ireland,* 101.

38. T. C. Barnard, "Planters and Policies in Cromwellian Ireland," *Past and Present* 61 (1973): 32.

39. Clarke, "Colonisation," 153.

40. Foster, *Modern Ireland,* 85; J. Bardon, *A History of Ulster,* 2nd ed. (Belfast: Blackstaff Press, 2001), 137–38.

41. N. Canny, *Making Ireland British 1580–1650* (Oxford: Oxford University Press, 2005), 555; P. J. Corish, "The Rising of 1641 and the Catholic Confederacy, 1641–45," in Moody, Martin, and Byrne, *A New History of Ireland III,* 292.

42. Foster, *Modern Ireland,* 101.

43. Ibid., 86.

44. Gillespie, "The Native Irish."

45. P. J. Corish, "The Cromwellian Regime, 1650–60," in Moody, Martin, and Byrne, *A New History of Ireland III,* 358.

46. Ibid., 374.

47. Foster, *Modern Ireland,* 111.

48. Beckett, *Making of Modern Ireland,* 109.

49. F. S. L. Lyons, *Culture and Anarchy in Ireland* (Oxford: Clarendon Press, 1980), 192.

3. Religion and Society in Pre-Famine Ireland

1. D. W. Miller, "Irish Catholicism and the Great Famine," *Journal of Social History* 9 (1975): 83.

2. S. J. Connolly, "Mass Politics and Sectarian Conflict, 1823–30," in *A New History of Ireland V: Ireland under the Union I, 1801–70,* ed. W. E. Vaughan (Oxford: Oxford University Press, 1989), 78.

3. Miller, "Irish Catholicism," 95.

4. M. MacCarthy-Morrogh, *The Munster Plantation: English Migration to Southern Ireland 1583–1641* (Oxford: Clarendon Press, 1986), 276.

5. A. Clarke, "Plantation and the Catholic Question, 1603–23," in *A New History of Ireland III: Early Modern Ireland 1534–1691,* ed. T. W. Moody, F. X. Martin, and F. J. Byrne (Oxford: Clarendon Press, 1976), 223.

6. J. L. McCracken, "The Social Structure and Social Life, 1714–60," in *A New History of Ireland IV: Eighteenth Century Ireland 1691–1800,* ed. T. W. Moody and W. E. Vaughan (Oxford: Oxford University Press, 1986), 31.

7. R. F. Foster, *Modern Ireland 1600–1972* (Harmondsworth: Penguin, 1989), 7.

8. N. Canny, "Early Modern Ireland c. 1500–1700," in *The Oxford Illustrated History of Ireland,* ed. R. F. Foster (Oxford: Oxford University Press, 1989), 135.

9. P. Robinson, "Plantation and Colonisation: The Historical Background," in *Integration and Division: Geographical Perspectives on the Northern Ireland Problem,* ed. F. W. Boal and J. N. H. Douglas (London: Academic Press, 1982), 31.

10. D. B. Quinn and K. W. Nicholls, "Ireland in 1534," in Moody, Martin, and Byrne, *A New History of Ireland III,* 33.

11. Foster, *Modern Ireland,* 73.

12. J. C. Beckett, *The Making of Modern Ireland 1603–1923* (London: Faber & Faber, 1966), 47.

13. A. T. Q. Stewart, *The Narrow Ground: Aspects of Ulster 1609–1969* (Belfast: Blackstaff, 1977), 35.

14. Foster, *Modern Ireland,* 323; C. Kinealy, *This Great Calamity: The Irish Famine 1845–52* (Dublin: Gill & Macmillan, 1995), 5–10.

15. These findings need to be treated with some caution due to the inconsistency in the sorts of geographical units being compared. For Ireland, the data are at the level of the baronies, which vary significantly in physical area but in 1831 had a mean area of 67,495 acres over 301 units. For England and Wales the closest comparable land divisions were the boroughs, cities, hundreds, wards, and wapentakes, which numbered over 1,000. These had a mean acreage of about 35,000.

16. Stewart, *Narrow Ground,* 129.

17. O. MacDonagh, "The Economy and Society, 1830–1845," in Vaughan, *A New History of Ireland V,* 229.

18. Beckett, *Making of Modern Ireland,* 291.

19. Ibid., 290.

20. C. Ó Gráda, *Ireland: A New Economic History 1780–1939* (Oxford: Clarendon Press, 1994), 74.

21. Ibid., 77–78.

22. M. Elliott, *The Catholics of Ulster* (London: Penguin, 2001), 313–14.

23. Ó Gráda, *Ireland,* 76–77.

24. Foster, *Modern Ireland,* 318–19.

25. J. J. Lee, *The Modernisation of Irish Society* (Dublin: Gill & Macmillan, 1989), 4.

26. Beckett, *Making of Modern Ireland,* 293.

27. Lee, *Modernisation,* 5.

28. D. H. Akenson, *Small Differences: Irish Catholics and Irish Protestants 1815–1922* (Montreal: McGill-Queen's University Press; Dublin: Gill & Macmillan, 1991), 26.

29. Quoted in Ó Gráda, *Ireland,* 71.

4. The Famine and Its Impacts, 1840s to 1860s

1. M. E. Daly, *The Slow Failure: Population Decline in Independent Ireland, 1920–1973* (Madison: University of Wisconsin Press, 2006), 3, 5.

2. Ibid., 5–6.

3. J. C. Beckett, *The Making of Modern Ireland 1603–1923* (London: Faber & Faber, 1966), 337.

4. S. J. Campbell, *The Great Irish Famine* (Strokestown: Famine Museum, 1994), 20.

5. R. F. Foster, *Modern Ireland 1600–1972* (Harmondsworth: Penguin, 1989), 319.

6. Ibid., 319–20.

7. J. Mokyr, *Why Ireland Starved: A Quantitative and Analytical History of the Irish Economy, 1800–1850* (London: George Allen & Unwin, 1985), 57.

8. Beckett, *Making of Modern Ireland,* 337.

9. E. R. R. Green, "The Great Famine 1845–50," in *The Course of Irish History,* ed. T. W. Moody and F. X. Martin, 4th ed. (Cork: Mercier Press, 2001), 224.

10. Ibid.

11. A. Jackson, *Ireland 1798–1998* (Oxford: Blackwell, 1999), 70.

12. C. Ó Gráda, *Ireland: A New Economic History 1780–1939* (Oxford: Clarendon Press, 1994), 177.

13. J. S. Donnelly Jr., "Excess Mortality and Emigration," in *A New History of Ireland V: Ireland under the Union I 1801–1870,* ed. W. E. Vaughan (Oxford: Clarendon Press, 1989), 351.

14. Jackson, *Ireland,* 69; Ó Gráda, *Ireland,* 173–74.

15. J. R. Donnelly Jr., "The Administration of Relief, 1847–51," in Vaughan, *A New History of Ireland V,* 331.

16. Jackson, *Ireland,* 69.

17. Foster, *Modern Ireland,* 318.

18. Ó Gráda, *Ireland,* 176.

19. Foster, *Modern Ireland,* 318.

20. L. M. Cullen, *An Economic History of Ireland since 1660* (London: Batsford, 1972), 132.

21. Donnelly, "Excess Mortality," 354.

22. S. H. Cousens, "The Regional Variation in Emigration from Ireland between 1821 and 1841," *Transactions of the Institute of British Geographers* 37 (1965): 27.

23. Foster, *Modern Ireland,* 318.

24. Mokyr, *Why Ireland Starved,* 267.

25. Donnelly, "Excess Mortality," 351.

26. L. Kennedy, P. S. Ell, E. M. Crawford, and L. A. Clarkson, *Mapping the Great Irish Famine* (Dublin: Four Courts Press, 1999), 18.

27. Campbell, *Great Irish Famine,* 33.

28. Donnelly, "Administration of Relief," 318; Campbell, *Great Irish Famine,* 33.

29. Foster, *Modern Ireland,* 316.

30. Jackson, *Ireland,* 56.

31. M. P. A. Macourt, "Religion as a Key Variable for Northern Ireland," *Environment and Planning A* 27 (1995): 595.

32. J. Lee, *The Modernisation of Irish Society* (Dublin: Gill & Macmillan, 1989), 51.

33. D. Dorling and P. Rees, "A Nation Still Dividing: The British Census and Social Polarisation, 1971–2001," *Environment and Planning A* 35 (2003): 1289.

34. V. Robinson, "Lieberson's Isolation Index: A Case Study Evaluation," *Area* 12 (1980): 307–12. This paper also compares the isolation index to other measures of segregation.

35. D. H. Akenson, *Small Differences: Irish Catholics and Irish Protestants 1815–1922* (Montreal: McGill-Queen's University Press; Dublin: Gill & Macmillan, 1991), 144.

36. J. Hart, "Sir Charles Trevelyan at the Treasury," *English Historical Review* 75, no. 294 (1960): 99.

37. R. Haines, *Charles Trevelyan and the Great Irish Famine* (Dublin: Four Courts Press, 2004), 4, 9–17.

38. Correspondence, reports, printed material, etc.: Civil Service Reform—Note on "night refuges," 25 July 1976; and C. E. Trevelyan, *The No-Popery Agitation and the Liverpool Corporation Schools,* 4th ed. (London: James Ridgeway, 1840), 1–16, both in CET 55/1, Charles Edward Trevelyan Papers, Robinson Library, Newcastle University.

39. R. Dunlop, "The Famine Crisis: Theological Interpretations and Implications," in *"Fearful*

Realities": New Perspectives on the Famine, ed. C. Morash and R. Hayes (Dublin: Irish Academic Press, 1996), 167.

40. Quoted in Ó Gráda, *Ireland,* 236.

41. Mokyr, *Why Ireland Starved,* 50–51; P. M. Solar, "The Great Famine Was No Ordinary Subsistence Crisis," in *Famine: The Irish Experience 900–1900,* ed. E. M. Crawford (Edinburgh: John Donald Publishers Ltd., 1989), 128–29.

42. Ó Gráda, *Ireland,* 236.

43. Kennedy et al., *Mapping the Famine,* 76.

44. Ó Gráda, *Ireland,* 187.

45. Lee, *Modernisation,* 2.

46. C. Ó Gráda, "Poverty, Population, and Agriculture, 1801–45," in Vaughan, *A New History of Ireland V,* 113.

47. Ibid.

48. Beckett, *Making of Modern Ireland,* 352–53.

49. Foster, *Modern Ireland,* 322.

50. Green, "Great Famine," 222.

51. J. Bardon, *A History of Ulster,* 2nd ed. (Belfast: Blackstaff Press, 2001), 320.

52. R. V. Comerford, "Ireland 1850–70: Post-famine and Mid-Victorian," in Vaughan, *A New History of Ireland V,* 381.

53. Foster, *Modern Ireland,* 336.

54. Comerford, "Post-famine and Mid-Victorian," 380–81.

55. Lee, *Modernisation,* 3–4.

56. Ibid., 10.

57. J. S. Donnelly Jr., "Landlords and Tenants," in Vaughan, *A New History of Ireland V,* 337.

58. Foster, *Modern Ireland,* 336.

59. Ó Gráda, *Ireland,* 125–26.

60. D. G. Boyce, *Ireland 1828–1923: From Ascendancy to Democracy* (Oxford: Blackwell, 1992), 37.

61. Bardon, *History of Ulster,* 185.

62. This development of mills in the Belfast region was facilitated by the greater availability of capital there than possibly anywhere else in the United Kingdom. This helps to underline how Ulster's longer-term prosperity, based in linen, acted to stimulate further expansion and advances in Belfast and the Lagan valley. See F. Geary, "The Rise and Fall of the Belfast Cotton Industry: Some Problems," *Irish Social and Economic History* 8 (1981): 30–49.

63. J. C. Beckett et al., *Belfast—the Making of the City* (Belfast: Appletree Press, 1983), 17; Bardon, *History of Ulster,* 334.

64. C. Ó Gráda, "Industry and Communications, 1801–45," in Vaughan, *A New History of Ireland V,* 146.

65. Ó Gráda, *Ireland,* 137.

66. Ó Gráda, "Industry and Communications," 146.

67. Ó Gráda, *Ireland,* 271.

68. Bardon, *History of Ulster,* 338.

69. Ibid., 299.

5. Toward Partition, 1860s to 1910s

1. M. E. Daly, *The Slow Failure: Population Decline in Independent Ireland, 1920–1973* (Madison: University of Wisconsin Press, 2006), 1.

2. L. Kennedy, P. S. Ell, E. M. Crawford, and L. A. Clarkson, *Mapping the Great Irish Famine* (Dublin: Four Courts Press, 1999), 18.

3. J. Bardon, *A History of Ulster,* 2nd ed. (Belfast: Blackstaff Press, 2001), 326.

4. C. Ó Gráda, *Ireland: A New Economic History 1780–1939* (Oxford: Clarendon Press, 1994), 291.

5. Bardon, *History of Ulster,* 328–29; for the making of handkerchiefs, see H. D. Gribbon, "Economic and Social History, 1850–1921," in *A New History of Ireland VI: Ireland under the Union II 1870–1921,* ed. W. E. Vaughan (Oxford: Clarendon Press, 1996), 292–93.

6. Bardon, *History of Ulster,* 327.

7. Ibid., 288.

8. Ibid., 294.

9. F. S. L. Lyons, *Ireland since the Famine,* 2nd ed. (London: Fontana Press, 1985), 66.

10. Bardon, *History of Ulster,* 390–91.

11. Ibid., 391; R. F. Foster, *Modern Ireland 1600–1972* (Harmondsworth: Penguin, 1989), 388.

12. Bardon, *History of Ulster,* 336.

13. S. Mountfield, *Western Gateway: A History of the Mersey Docks and Harbour Board* (Liverpool: Liverpool University Press, 1965), 203–205.

14. Gribbon, "Economic and Social History," 294.

15. Bardon, *History of Ulster,* 332–33.

16. L. P. Curtis Jr., "Ireland in 1914," in Vaughan, *A New History of Ireland VI,* 165. Bardon, *History of Ulster,* 327.

17. Gribbon, "Economic and Social History," 301.

18. Ó Gráda, *Ireland,* 304.

19. Ibid., 298.

20. E. Malcolm, "Temperance and Irish Nationalism," in *Ireland under the Union: Varieties of Tension,* ed. F. S. L. Lyons and R. A. J. Hawkins (Oxford: Clarendon Press, 1980), 70, 74–75.

21. Foster, *Modern Ireland,* 321–22.

22. F. S. L. Lyons, *Culture and Anarchy in Ireland 1890–1939* (Oxford: Clarendon Press, 1979), 19.

23. MacNeice quoted in D. Kiberd, *Inventing Ireland: The Literature of the Modern Nation* (London: Jonathan Cape, 1995), 449.

24. J. Lee, *The Modernisation of Irish Society* (Dublin: Gill & Macmillan, 1989), 9–10.

25. Ibid., 15. The underlying historically deterministic assumptions of the Weberian "Protestant ethic" are discussed in H. Akenson, *Small Differences: Irish Catholics and Irish Protestants 1815–1922* (Montreal: McGill-Queen's University Press; Dublin: Gill & Macmillan, 1991), 16. They provided contemporary commentators with a convenient tool for explaining the structural inequalities in the Irish economy in religious terms.

26. R. Sweetman, "The Development of the Port," in *Belfast—the Making of the City,* by J. C. Beckett et al. (Belfast: Appletree Press, 1983), 61–62.

27. Ó Gráda, *Ireland,* 297.

28. D. Fitzpatrick, *Irish Emigration 1801–1921* (Dundalk: Dundalgan Press, 1984), 30.

29. Ibid., 649.

30. Bardon, *History of Ulster,* 337; Ó Gráda, *Ireland,* 291.

31. D. Fitzpatrick, "Emigration, 1871–1921," in Vaughan, *A New History of Ireland VI,* 612.

32. Ibid., 610.

33. Ó Gráda, *Ireland,* 226.

34. Fitzpatrick, "Emigration," 612.

35. R. C. Geary, "The Future Population of Saorstát Éireann and Some Observations on Population Statistics," *Journal of the Statistical and Social Inquiry Society of Ireland* 15, no. 6 (1936): 24.

36. Ibid.

37. Ibid.

38. Quoted in Lee, *Modernisation,* 8.

39. Foster, *Modern Ireland,* 351.

40. M. Turner, *After the Famine: Irish Agriculture, 1850–1914* (Cambridge: Cambridge University Press, 1996), 164.

41. R. V. Comerford, "The Parnell Era, 1883–91," in Vaughan, *A New History of Ireland VI,* 73.

42. Gribbon, "Economic and Social History," 283.

43. T. J. Hatton and J. G. Williamson, "After the Famine: Emigration from Ireland, 1850–1913," *Journal of Economic History* 53 (1993): 595.

44. Note that this is slightly different from the index of dissimilarity for the same data calculated using the diocese. This is because the arrangement of the spatial units will affect the results.

45. T. W. Moody, "Fenianism, Home Rule and the Land War," in *The Course of Irish History,* ed. T. W. Moody and F. X. Martin, 4th ed. (Cork: Mercier Press, 2001), 236–37.

46. Ibid.

47. R. V. Comerford, "The Politics of Distress, 1877–82," in Vaughan, *A New History of Ireland VI,* 35.

48. R. Kee, *The Laurel and the Ivy* (Harmondsworth: Penguin, 1993), 184.

49. D. Hempton and M. Hill, *Evangelical Protestantism in Ulster Society 1740–1890* (London: Routledge, 1992), 166; E. R. Norman, *The Catholic Church and Ireland in the Age of Rebellion: 1859–1873* (London: Longmans, 1965), 20–21.

50. Turner, *After the Famine,* 210.

51. P. Bew, *Conflict and Conciliation in Ireland 1890–1910: Parnellites and Radical Agrarians* (Oxford: Clarendon Press, 1987), 17.

52. P. Bew, *Charles Stewart Parnell* (Dublin: Gill & Macmillan, 1980), 70; D. Fitzpatrick, "The Geography of Irish Nationalism, 1910–1921," in *Nationalism and Popular Protest in Ireland,* ed. C. H. E. Philpin (Cambridge: Cambridge University Press, 1987), 425.

53. 207 Parl. Deb., H.C. (12 July 1871) 1542 (U.K.).

54. D. G. Boyce, *Ireland 1828–1923: From Ascendancy to Democracy* (Oxford: Blackwell, 1992), 57.

55. Kiberd, *Inventing Ireland,* 25.

56. A. C. Hepburn, *The Conflict of Nationality in Modern Ireland* (London: Edward Arnold, 1980), 36.

57. A. Boyd, *Holy War in Belfast,* 3rd ed. (Belfast: Pretani Press, 1987), 19–20.

58. E. Larkin, "The Devotional Revolution in Ireland, 1850–75," *American Historical Review* 77 (1972): 644; K. T. Hoppen, *Elections, Politics and Society in Ireland, 1832–1885* (Oxford: Clarendon Press, 1985), 171.

59. D. Bowen, *Paul Cardinal Cullen and the Shaping of Modern Irish Catholicism* (Dublin: Gill & Macmillan, 1983), 131.

60. Bardon, *History of Ulster,* 350.

61. M. Elliott, *The Catholics of Ulster* (London: Penguin, 2001), 357.

62. R. Douglas, L. Harte, and J. O'Hara, *Drawing Conclusions: A Cartoon History of Anglo-Irish Relations 1978–1998* (Belfast: Blackstaff Press, 1998), 99.

63. 38 Parl. Deb., H.C. (10 March 1896) 573 (U.K.).

64. W. A. Maguire, *Belfast* (Keele: Ryburn Publishing, 1993), 102.

65. Ibid.

66. Bardon, *History of Ulster,* 405.

6. Partition and Civil War, 1911 to 1926

1. M. E. Daly, *The Slow Failure: Population Decline in Independent Ireland, 1920–1973* (Madison: University of Wisconsin Press, 2006), 6.

2. Ibid., 7.

3. B. Collins, "The Edwardian City," in *Belfast: The Making of the City,* by J. C. Beckett et al. (Belfast: Appletree Press, 1983), 167.

4. L. Clarkson, "The City and the Country," in Beckett et al., *Belfast,* 161.

5. A. J. Christopher, "'The Second City of the Empire': Colonial Dublin, 1911," *Journal of Historical Geography* 23 (1997): 151.

6. The quote is the final line of stanzas 1, 2, and 4 of W. B. Yeats's poem "Easter, 1916," published in 1920.

7. M. Laffan, *The Partition of Ireland 1911–1925* (Dundalk: Dundalgan Press, 1983), 1.

8. J. J. Lee, *Ireland 1912–1985: Politics and Society* (Cambridge: Cambridge University Press, 1989), 17–18.

9. A. Jackson, *Ireland 1798–1998* (Oxford: Blackwell, 1999), 232.

10. A. T. Q. Stewart, *The Narrow Ground: Aspects of Ulster 1609–1969* (Belfast: Blackstaff, 1977), 168.

11. Ibid.

12. C. Townsend, *Political Violence in Ireland: Government and Resistance since 1848* (Oxford: Clarendon Press, 1983), 268.

13. J. C. Beckett, *The Making of Modern Ireland 1603–1923* (London: Faber & Faber, 1966), 430–31.

14. D. G. Boyce, *Ireland 1828–1923: From Ascendancy to Democracy* (Oxford: Blackwell, 1992), 80.

15. N. Mansergh, *The Unresolved Question: The Anglo-Irish Settlement and Its Undoing 1912–72* (New Haven, Conn.: Yale University Press, 1991), 80–81.

16. Laffan, *Partition,* 47.

17. J. Bardon, *A History of Ulster,* 2nd ed. (Belfast: Blackstaff Press, 2001), 449.

18. A. C. Hepburn, *The Conflict of Nationality in Modern Ireland* (London: Edward Arnold, 1980), 84.

19. R. Douglas, L. Harte, and J. O'Hara, *Drawing Conclusions: A Cartoon History of Anglo-Irish Relations 1978–1998* (Belfast: Blackstaff Press, 1998), 176.

20. R. F. Foster, *Modern Ireland 1600–1972* (Harmondsworth: Penguin, 1989), 473.

21. Lee, *Ireland 1912–1985,* 22.

22. Bardon, *History of Ulster,* 455.

23. F. S. L. Lyons, "The Rising and After," in *A New History of Ireland VI: Ireland under the Union II 1870–1921,* ed. W. E. Vaughan (Oxford: Clarendon Press, 1996), 222.

24. R. Dudley Edwards, *Patrick Pearse: The Triumph of Failure* (London: Victor Gollancz, 1977), 285.

25. M. Laffan, *The Resurrection of Ireland: The Sinn Féin Party 1916–1923* (Cambridge: Cambridge University Press, 1999), 49–51.

26. Lee, *Ireland 1912–1985,* 40. There was a geography to political activism during the revolutionary period, with Sinn Féin membership tending to be lower in the counties of the eastern seaboard and lowest of all in Dublin. See D. Fitzpatrick, "The Geography of Irish Nationalism," *Past and Present* 78 (1978): 144.

27. Lee, *Ireland 1912–1985,* 42.

28. F. S. L. Lyons, *Ireland since the Famine,* 2nd ed. (London: Fontana Press, 1985), 408, 413–14.

29. Foster, *Modern Ireland,* 501.

30. Hepburn, *Conflict of Nationality,* 123.

31. Lee, *Ireland 1912–1985,* 54; Earl of Longford and T. P. O'Neill, *Eamon de Valera* (London: Hutchinson & Co., 1970), 175.

32. T. P. Coogan and G. Morrison, *The Irish Civil War* (London: Weidenfeld & Nicolson, 1998), 54.

33. E. O'Halpin, *Defending Ireland: The Irish State and Its Enemies since 1922* (Oxford: Oxford University Press, 1999), 26.

34. Foster, *Modern Ireland,* 503.

35. P. Buckland, *A History of Northern Ireland* (Dublin: Gill & Macmillan, 1981), 1; Lee, *Ireland 1912–1985,* 44–45.

36. F. Pakenham, *Peace by Ordeal: The Negotiation of the Anglo-Irish Treaty, 1921* (London: Pimlico, 1992), 289.

37. C. Costello, *A Most Delightful Station: The British Army on the Curragh of Kildare* (Ann Arbor, Mich.: Collins Press, 1996), 327–31.

38. The "U" stands for "Urban."

39. Lee, *Ireland 1912–1985,* 59–61.

40. P. Hart, *The I.R.A. and Its Enemies: Violence and Community in Cork, 1916–1923* (Oxford: Oxford University Press, 1999), 277.

41. See representative reports in the *Irish Times,* 3, 4, 5, and 6 May 1922.

42. Lee, *Ireland 1912–1985,* 60.

43. M. Elliott, *The Catholics of Ulster* (London: Penguin, 2001), 374.

44. Quoted in ibid.

45. Foster, *Modern Ireland,* 533.

46. Elliott, *Catholics,* 379.

47. Buckland, *History of Northern Ireland,* 41. The USC was almost completely Protestant in its composition, and little effort was made to encourage Catholics to join the force. The reservist B section of the USC, which would become better known as the B Specials, enjoyed a particularly poor rapport with the Catholic community. They were disbanded in 1969 as a response to their role in that year's summer of intense violence in Northern Ireland, which marked the start of the modern Troubles.

48. Laffan, *Partition,* 1.

49. Buckland, *History of Northern Ireland,* 20.

50. Foster, *Modern Ireland,* 443.

51. Pakenham, *Peace by Ordeal,* 221.

52. K. J. Rankin, "The Creation and Consolidation of the Irish Border," working paper no. 48, p. 24, Institute of British-Irish Studies, University College Dublin, 2005; *Report of the Irish Boundary Commission* (Shannon: Irish Academic Press, 1969), xvii–xviii.

53. N. Mansergh, "The Government of Ireland Act, 1920," in *Historical Studies IX,* ed. J. G. Barry (Belfast: Blackstaff Press, 1974), 74.

54. Ibid., 66.

55. Lee, *Ireland 1912–1985,* 45.

56. Laffan, *Partition,* 68.

57. H. Patterson, "In the Land of King Canute: The Influence of Border Unionism on Ulster Unionist Politics, 1945–63," *Contemporary British History* 20, no. 4 (2006): 511–12.

58. Stewart, *Narrow Ground,* 47–48.

59. K. J. Rankin, "The Role of the Irish Boundary Commission in the Entrenchment of the Irish Border: From Tactical Panacea to Political Liability," *Journal of Historical Geography* 34 (2008): 446.

7. Division and Continuity, 1920s to 1960s

1. B. Anderson, *Imagined Communities,* 2nd ed. (London: Verso, 2006), 7.

2. M. E. Daly, *The Slow Failure: Population Decline in Independent Ireland, 1920–1973* (Madison: University of Wisconsin Press, 2006), 9.

3. H. Akenson, *Small Differences: Irish Catholics and Irish Protestants 1815–1922* (Montreal: McGill-Queen's University Press; Dublin: Gill & Macmillan, 1991), 16–19; J. Lee, *The Modernisation of Irish Society* (Dublin: Gill & Macmillan, 1989), 16.

4. J. Bardon, *A History of Ulster,* 2nd ed. (Belfast: Blackstaff Press, 2001), 400.

5. R. Douglas, L. Harte, and J. O'Hara, *Drawing Conclusions: A Cartoon History of Anglo-Irish Relations 1978–1998* (Belfast: Blackstaff Press, 1998), 240; J. Meehan, "Preliminary Notes on the Census of Population in Northern Ireland, 1937," *Journal of the Statistical and Social Inquiry Society of Ireland* 15 (1937): 78.

6. Daly, *Slow Failure,* 9.

7. J. J. Lee, *Ireland 1912–1985: Politics and Society* (Cambridge: Cambridge University Press, 1989), 237.

8. R. Fanning, *Independent Ireland* (Dublin: Helicon, 1983), 102–103.

9. Douglas, Harte, and O'Hara, *Drawing Conclusions,* 202.

10. T. Brown, *Ireland: A Social and Cultural History* (London: Fontana Press, 1981), 15.

11. D. Keogh, *Twentieth-Century Ireland: Nation and State* (Dublin: Gill & Macmillan, 1994), 36.

12. M. E. Daly, *Industrial Development and Irish National Identity* (Syracuse, N.Y.: Syracuse University Press, 1992), 16.

13. Daly, *Slow Failure,* 30–31.

14. Ibid., 23.

15. Ibid., 26.

16. L. M. Cullen, *An Economic History of Ireland since 1660* (London: B. T. Batsford, 1972), 175.

17. Daly, *Slow Failure,* 22.

18. C. Ó Gráda, *Ireland: A New Economic History 1780–1939* (Oxford: Clarendon Press, 1994), 409–10.

19. P. Lynch, "The Irish Free State and the Republic of Ireland: 1921–66," in *The Course of Irish History,* ed. T. W. Moody and F. X. Martin, 4th ed. (Cork: Mercier Press, 2001), 282.

20. Daly, *Slow Failure,* 45.

21. D. S. Johnson and L. Kennedy, "The Two Economies in Ireland in the Twentieth Century," in *A New History of Ireland Vol. VII: Ireland 1921–1984,* ed. J. R. Hill (Oxford: Oxford University Press, 2003), 453.

22. Keogh, *Twentieth-Century Ireland,* 67.

23. D. S. Johnson, "The Economic History of Ireland between the Wars," *Irish Economic and Social History* 1 (1974): 57.

24. Lee, *Ireland 1912–1985,* 187; Keogh, *Twentieth-Century Ireland,* 89.

25. F. S. L. Lyons, *Ireland since the Famine,* 2nd ed. (London: Fontana Press, 1985), 614.

26. Brown, *Ireland,* 144.

27. P. Buckland, *The Factory of Grievances: Devolved Government in Northern Ireland 1921–39* (Dublin: Gill & Macmillan, 1979), 2.

28. Johnson and Kennedy, "The Two Economies," 465–67; R. F. Foster, *Modern Ireland 1600–1972* (Harmondsworth: Penguin, 1989), 528.

29. Bardon, *History of Ulster,* 529.

30. Ibid.

31. Northern Ireland Statistics and Research Agency, "Census Taking in Ireland," http://www.nisranew.nisra.gov.uk/census/censushistory/censusireland.html, accessed 11 September 2009.

32. Buckland, *Factory of Grievances,* 5–6.

33. *Bunreacht na hÉireann/Constitution of Ireland* (Dublin: Stationery Office, 1937), 4.

34. Lee, *Ireland 1912–1985,* 264; Keogh, *Twentieth-Century Ireland,* 109.

35. Lee, *Ireland 1912–1985,* 256.

36. Ibid., 232.

37. J. J. Sexton, "Emigration and Immigration in the Twentieth Century," in Hill, *A New History of Ireland Vol. VII,* 801.

38. Lee, *Ireland 1912–1985,* 384.

39. Brown, *Ireland,* 108–109.

40. Ibid., 109.

41. E. Bowen, *The Last September* (London: Vintage, 1998), 147.

42. F. S. L. Lyons, "The Minority Problem in the 26 Counties," in *The Years of the Great Test 1926–39,* ed. F. MacManus (Cork: Mercier Press, 1967), 97.

43. Foster, *Modern Ireland,* 534.

44. J. H. Whyte, *Church and State in Modern Ireland, 1923–1970* (Dublin: Gill & Macmillan, 1971), 36–37.

45. Keogh, *Twentieth-Century Ireland,* 56.

46. Ibid., 56–57.

47. E. MacDermott, *Clann na Poblachta* (Cork: Cork University Press, 1998), 148–52.

48. D. Keogh, *Ireland and the Vatican* (Cork: Cork University Press, 1995), 133.

49. Lee, *Ireland 1912–1985,* 203.

50. Lyons, *Ireland since the Famine,* 473.

51. D. Miller, *Queen's Rebels: Ulster Loyalism in Historical Perspective* (Dublin: Gill & Macmillan, 1978), 136.

52. Douglas, Harte, and O'Hara, *Drawing Conclusions,* 223. The gerrymandering of Derry's boundaries allowed 7,444 unionists to return twelve councillors while 9,961 nationalists could elect just eight. See Foster, *Modern Ireland,* 557.

53. Douglas, Harte, and O'Hara, *Drawing Conclusions,* 223.

54. M. Connolly, *Politics and Policy-Making in Northern Ireland* (Basingstoke: Palgrave Macmillan, 1990), 113.

55. Bardon, *History of Ulster,* 502; Buckland, *History of Northern Ireland,* 57.

56. Bardon, *History of Ulster,* 502; Buckland, *History of Northern Ireland,* 77.

57. Buckland, *History of Northern Ireland,* 77.

58. W. B. Yeats, "Remorse for Intemperate Speech," in *Selected Poetry* (London: Penguin, 1991), 179.

59. J. H. Whyte, "To the Declaration of the Republic and the Ireland Act, 1945–49," in Hill, *A New History of Ireland Vol. VII,* 275.

60. Ibid., 276.

8. Toward the Celtic Tiger: The Republic, 1961 to 2002

1. J. J. Lee, *Ireland 1912–1985: Politics and Society* (Cambridge: Cambridge University Press, 1989), 334–35.

2. J. Blackwell, "Government, Economy and Society," in *Unequal Achievement: The Irish Experience 1957–1982,* ed. F. Litton (Dublin: Institute of Public Administration, 1982), 43.

3. L. Kennedy, *The Modern Industrialisation of Ireland, 1940–1988,* Studies in Irish Economic and Social History 5 (Dublin: Economic and Social History Society of Ireland, 1989), 13.

4. J. Lee, "Society and Culture," in Litton, *Unequal Achievement,* 6.

5. D. Ferriter, *The Transformation of Ireland 1900–2000* (London: Profile Books, 2004), 469.

6. Kennedy, *Modern Industrialisation,* 17.

7. R. F. Foster, *Modern Ireland 1600–1972* (Harmondsworth: Penguin, 1989), 577–78.

8. Ibid., 580.

9. J. J. Sexton, "Emigration and Immigration in the Twentieth Century," in *A New History of Ireland Vol. VII: Ireland 1921–1984,* ed. J. R. Hill (Oxford: Oxford University Press, 2003), 810. Sexton's figures show that an estimated 436,000 left the Republic of Ireland between 1981 and 1991 (ibid., 802). This led to a fall in the overall population between 1987 and 1991, but the overall trend of net growth between major censuses (1981 and 1991) was maintained (ibid., 797).

10. D. Keogh, *Twentieth-Century Ireland: Nation and State* (Dublin: Gill & Macmillan, 1994), 244.

11. P. Bew and H. Patterson, *Seán Lemass and the Making of Modern Ireland 1945–66* (Dublin: Gill & Macmillan, 1982), 160.

12. M. E. Daly, *The Slow Failure: Population Decline in Independent Ireland, 1920–1973* (Madison: University of Wisconsin Press, 2006), 247.

13. Ibid.; V. Buckley, *Memory Ireland: Insights into the Contemporary Irish Condition* (Harmondsworth: Penguin, 1985), 40.

14. L. Bardwell, "Tallaght II," in *Invisible Dublin: A Journey through the City's Suburbs,* ed. D. Bolger (Dublin: Raven Arts Press, 1991), 139–40.

15. N. Scheper-Hughes, *Saints, Scholars and Schizophrenics* (Berkeley: University of California Press, 1979), 37–39.

16. J. Healy, *"Nobody Shouted Stop": The Death of an Irish Town* (Cork: Mercier Press, 1968); J. Waters, *Jiving at the Crossroads* (Belfast: Blackstaff Press, 1991).

17. Keogh, *Twentieth-Century Ireland,* 243.

18. K. A. Kennedy, T. Giblin, and D. McHugh, *The Economic Development of Ireland in the Twentieth Century* (London: Routledge, 1988), 62–63.

19. J. H. Whyte, "Economic Progress and Political Pragmatism, 1957–63," in Hill, *A New History of Ireland Vol. VII,* 295; J. K. Jacobsen, *Chasing Progress in the Irish Republic* (Cambridge: Cambridge University Press, 1994), 79.

20. Colin Buchanan and Partners, *Regional Development in Ireland* (Dublin: An Foras Forbartha, 1968).

21. D. F. Hannan and P. Commins, "The Significance of Small-Scale Landowners," in *The Development of Industrial Society in Ireland,* ed. J. H. Goldthorpe and C. T. Whelan, Proceed-

ings of the British Academy 79 (Oxford: British Academy, 1990), 99.

22. P. Mair, *The Changing Irish Party System* (London: Pinter Publishers, 1987), 138; J. Joyce and P. Murtagh, *The Boss: Charles J. Haughey in Government* (Swords: Poolbeg Press, 1983), 351–54.

23. E. Hazelkorn, "Class, Clientelism and the Political Process," in *Ireland: A Sociological Profile,* ed. P. Clancy, S. Drury, K. Lynch, and L. O'Dowd (Dublin: Institute of Public Administration, 1986), 338.

24. Waters, *Jiving at the Crossroads,* 26.

25. Daly, *Slow Failure,* 255.

26. Ibid.; Ferriter, *Transformation,* 469.

27. J. Ruane, "Success and Failure in a West of Ireland Factory," in *Ireland from Below: Social Change and Local Communities,* ed. C. Curtin and T. M. Wilson (Galway: Galway University Press, 1987), 181–84.

28. P. Breathnach, "Exploring the 'Celtic Tiger' Phenomenon: Causes and Consequences of Ireland's Economic Miracle," *European Urban and Regional Studies* 5 (1998): 305–16.

29. J. Coolahan, "Higher Education, 1908–84," in Hill, *A New History of Ireland Vol. VII,* 785–87.

30. Quoted in J. O'Malley, "Campaigns, Manifestoes and Party Finances," in *Ireland at the Polls 1981, 1982 and 1987: A Study of Four General Elections,* ed. H. Penniman and B. Farrell (Durham, N.C.: Duke University Press, 1987), 33.

31. Coolahan, "Higher Education," 789.

32. L. Gibbons, *Transformations in Irish Culture,* Critical Conditions: Field Day Essays and Monographs 2 (Cork: Cork University Press, 1996), 86–89.

33. Ferriter, *Transformation,* 662–63.

34. U. Kockel, "Countercultural Migrants in the West of Ireland," in *Contemporary Irish Migration,* ed. R. King (Dublin: Geographical Society of Ireland, 1991), 70–71.

35. J. Gleeson, R. Kitchin, B. Bartley, J. Driscoll, R. Foley, S. Fotheringham, and C. Lloyd, *The Atlas of the Island of Ireland: Mapping Social and Economic Change* (Maynooth/Armagh: AIRO/ICLRD, 2008), 49.

36. M. P. A. Macourt, "Using Census Data: Religion as a Key Variable in Studies of Northern Ireland," *Environment and Planning A* 27 (1995): 605. This particularly applies to Northern Ireland in recent years, but it should be stated that the Republic's census has also always alluded to this potential vulnerability in its collation procedure. See, for example, the explanatory notes to the religion question in the 1961 census, which states that enumerators should not press respondents for answers to the religion question. The same wording also appears in previous and subsequent years. See Central Statistics Office, "Census 1961 Vol. 7—Religion and Birthplaces—Explanatory Notes," http://www.cso.ie/census/census_1961_results/Volume7/C%201961%20VOL%207%20explan.pdf, accessed 4 September 2009.

37. M. Nic Ghiolla Phádraig, "Religious Practice and Secularisation," in Clancy et al., *Ireland,* 153.

38. T. Inglis, *Moral Monopoly: The Rise and Fall of the Catholic Church in Modern Ireland* (Dublin: University College Dublin Press, 1998), 204.

39. J. Coakley and M. Gallagher, *Politics in the Republic of Ireland,* 3rd ed. (Abingdon: Routledge, 1999), 46.

40. Ferriter, *Transformation,* 703; R. Breen and C. T. Whelan, "Social Mobility in Ireland: A Comparative Analysis," in *Ireland North and South: Perspectives from the Social Sciences,* ed. A. F. Heath, R. Breen, and C. T. Whelan, Proceedings of the British Academy 98 (Oxford: British Academy, 1999), 337; P. Lyons, "The Distribution of Personal Wealth in Ireland," in *Ireland: Some Problems of a Developing Economy,* ed. A. A. Tait and J. A. Bristow (Dublin: Gill & Macmillan, 1972), 178.

41. Foster, *Modern Ireland,* 596; Buckley, *Memory Ireland,* ix.

9. Stagnation and Segregation

1. I. Shuttleworth and C. Lloyd, "Are Northern Ireland's Communities Dividing? Evidence from Geographically Consistent Population Data 1971–2001," *Environment and Planning A* 41 (2009): 213–29.

2. These measurements refer to the length of the sides of the cells; thus, the 1 km cells have an area of 1 km^2, while the 100 m cells have an area of $100 \times 100 = 10,000$ m^2.

3. P. Compton and J. Power, "Estimates of the Religious Composition of Northern Ireland Local Government Districts in 1981 and Change in the Geographical Pattern of Religious Composition between 1971 and 1981," *Economic and Social Review* 17 (1986): 87–106.

4. Northern Ireland Statistics and Research Agency, "Population Statistics," http://www.nisra.gov.uk/demography/default.asp3.htm, accessed 27 March 2010.

5. A. G. Champion, *Counterurbanization: The Changing Pace & Nature of Population Deconcentration* (London: Edward Arnold, 1989).

6. P. Compton, *Demographic Review Northern Ireland 1995* (Belfast: Northern Ireland Economic Development Office, 1995).

7. *Registrar General's Annual Report* (Belfast: NISRA, 2009).

8. J. Smyth and A. Cebulla, "Belfast—Industrial Collapse and Post-Fordist Overdetermination," in *Development Ireland,* ed. P. Shirlow (London: Pluto, 1995), 81–93.

9. R. Osborne and I. Shuttleworth, "Fair Employment in Northern Ireland," in *Fair Employment in Northern Ireland: A Generation On,* ed. R. Osborne and I. Shuttleworth (Belfast: Blackstaff Press, 2004), 1–23.

10. J. Anderson and I. Shuttleworth, "Sectarian Demography, Territoriality and Policy in Northern Ireland," *Political Geography* 17 (1998): 187–208.

11. O. McEldowney, J. Anderson, and I. G. Shuttleworth, "Sectarian Demography: Dubious Discourses of Demography in Ethno-national Conflict," in *Political Discourse and Conflict Resolution: Debating Peace in Northern Ireland,* ed. K. Hayward and C. O'Donnell (London: Routledge, 2000), 160–76.

12. M. P. A. Macourt, "Using Census Data: Religion as a Key Variable in Studies of Northern Ireland," *Environment and Planning A* 27 (1995): 593–614.

13. J. O'Farrell, "Apartheid," *New Statesman,* 28 November 2005, http://www.newstatesman.com/200511280006, accessed 21 September 2011. See also McEldowney, Anderson, and Shuttleworth, "Sectarian Demography."

14. Shuttleworth and Lloyd, "Are Northern Ireland's Communities Dividing?," 213–29.

15. F. W. Boal, "Segregating and Mixing: Space and Residence in Belfast," in *Integration and Division: Geographical Perspectives on the Northern Ireland Problem,* ed. F. W. Boal and J. N. H. Douglas (London: Academic Press, 1982).

16. Shuttleworth and Lloyd, "Are Northern Ireland's Communities Dividing?," 213–29.

17. C. E. Brett, *Housing a Divided Community* (Dublin: Institute of Public Administration; Belfast: Institute of Irish Studies, Queen's University Belfast, 1986).

18. J. Power and I. Shuttleworth, "Intercensal Population Change in the Belfast Urban Area 1971–91: The Correlates of Population Increase and Decrease in a Divided Society," *International Journal of Population Geography* 3 (1997): 91–108.

19. Government of Northern Ireland, *Violence and Civil Disturbances in Northern Ireland in 1969: Report of Tribunal of Inquiry* (Belfast: HMSO, 1972), 248.

20. See D. Barritt and C. Carter, *The Northern Ireland Problem: A Study in Group Relations* (Oxford: Oxford University Press, 1962); R. J. Cormack and R. D. Osborne, eds., *Discrimination and Public Policy in Northern Ireland* (Oxford: Clarendon Press, 1991); E. McLaughlin and P. Quirk, *Policy Aspects of Employment Equality in Northern Ireland* (Belfast: Standing Advisory Commission on Human Rights, 1996).

10. Communal Conflict and Death in Northern Ireland, 1969 to 2001

1. B. O'Duffy and B. O'Leary, "Violence in Northern Ireland, 1969–June 1989," in *The Future of Northern Ireland,* ed. J. McGarry and B. O'Leary (Oxford: Clarendon Press, 1990), 319–21; M. T. Fay, M. Morrissey, and M. Smyth, *Mapping Troubles-Related Deaths in Northern Ireland 1969–1998,* 2nd ed. (Londonderry: INCORE [University of Ulster and United Nations University], 1998), 42–46.

2. P. Bishop and E. Mallie, *The Provisional IRA* (London: Corgi, 1988); T. P. Coogan, *The IRA,* 3rd ed. (London: Fontana Press, 1987); R. English, *Armed Struggle: The History of the IRA* (London: Pan, 2004); H. Patterson, *The Politics of Illusion: A Political History of the IRA,* 2nd ed. (London: Serif, 1997).

3. S. Bruce, *The Red Hand: Protestant Paramilitaries in Northern Ireland* (Oxford: Oxford University Press, 1992); M. Dillon, *The Shankill Butchers: A Case Study of Mass Murder* (London: Arrow, 1990); P. Taylor, *Loyalists* (London: Bloomsbury, 1999).

4. C. O'Neill, P. Durkin, D. McAlister, A. S. Dogra, and M. McAnespie, "In-Patient Costs and Paramilitary Punishment Beatings in Northern Ireland," *European Journal of Public Health* 12 (2002): 69–71.

5. http://cain.ulst.ac.uk/sutton/index.html, accessed 28 July 2011.

6. M. Sutton, *Bear in Mind These Dead: An Index of Deaths from the Conflict in Ireland 1969–1993* (Belfast: Beyond the Pale, 1994).

7. B. Lynn, "e-Resources on the Northern Ireland Conflict on CAIN," Irish Studies e-Resource Workshop, unpublished conference proceedings, 8 April 2011, Queen's University Belfast/Public Records Office of Northern Ireland (PRONI).

8. R. D. Sack, *Human Territoriality: Its Theory and History* (Cambridge: Cambridge University Press, 1986).

9. J. Anderson and I. Shuttleworth, "Sectarian Demography, Territoriality and Political Development in Northern Ireland," *Political Geography* 17 (1998): 187–208.

10. S. McGrellis, "Pure and Bitter Spaces: Gender, Identity and Territory in Northern Irish Youth Transitions," *Gender and Education* 17 (2005): 523–24.

11. C. Lloyd, *Local Models for Spatial Analysis* (Boca Raton, Fla.: CRC Press, 2006).

12. The Real IRA is a republican paramilitary group that rejected the IRA ceasefires of the mid- to late 1990s and has continued a campaign of violence. The INLA was the second largest republican paramilitary organization for much of the Troubles.

13. D. McKittrick, S. Kelters, B. Feeney, C. Thornton, and D. McVea, *Lost Lives: The Stories of the Men, Women and Children Who Died as a Result of the Northern Ireland Troubles,* 2nd ed. (Edinburgh: Mainstream Publishing, 2007), 1094–98.

14. P. Bew and G. Gillespie, *Northern Ireland: A Chronology of the Troubles, 1968–1999,* 2nd ed. (Dublin: Gill & Macmillan, 1999), 245, 291.

15. E. Moloney, *A Secret History of the IRA* (London: Penguin, 2002), 324.

16. Dillon, *Shankill Butchers,* 107.

17. D. McKittrick and D. McVea, *Making Sense of the Troubles,* 2nd ed. (London: Penguin, 2001), 70–72.

11. Belfast through the Troubles

1. I. Shuttleworth, P. Tyler, and D. McKinstry, "Job Loss, Training and Employability: What Can We Learn from the 2000 Harland and Wolff Redundancy?," *Environment and Planning A* 37 (2005): 1651–68.

2. A. G. Champion, *Counterurbanization: The Changing Pace & Nature of Population Deconcentration* (London: Edward Arnold, 1989). See also A. G. Champion, A. E. Green, D. W. Owen, D. J. Ellis, and M. G. Coombes, *Changing Places: Britain's Demographic, Economic & Social Complexion* (London: Edward Arnold, 1987).

3. It might be felt that the use of the community background definition of religion in 2001 could have affected these figures, but using the more directly comparable 1991 question also shows both of these clusters as being over 90 percent Catholic.

4. F. W. Boal and M. A. Poole, "Religious Residential Segregation in Belfast in Mid-1969: A Multi-level Analysis," *Social Patterns in Cities* (London: IBG Special Publication, 1973).

5. N. Bailey and M. Livingston, "Selective Migration and Neighbourhood Deprivation: Evidence from 2001 Census Migration Data for England and Scotland," *Urban Studies* 45 (2008): 943–61.

6. See D. Dorling and P. Rees, "A Nation Still Dividing: The British Census and Social Polarisation 1971–2001," *Environment and Planning A* 35 (2003): 1287–1313; and P. Breathnach, "Social Polarisation in the Post-Fordist Informational Economy: Ireland in International Context," *Irish Journal of Sociology* 11 (2002): 3–22.

12. Conclusions

1. D. Barritt and C. Carter, *The Northern Ireland Problem: A Study in Group Relations* (Oxford: Oxford University Press, 1962).

2. See R. B. McDowell, *Crisis and Decline: The Fate of the Southern Unionists* (Dublin: Lilliput Press, 1997), 136; F. S. L. Lyons, "The Minority Problem in the 26 Counties," in *The Years of the Great Test 1926–39,* ed. F. MacManus (Cork: Mercier Press, 1967), 97; T. Brown, *Ireland: A Social and Cultural History 1922–1985* (London: Fontana Press, 1985), 109; R. F. Foster, *Modern Ireland 1600–1972* (London: Penguin Books, 1989), 533–34; and F. S. L. Lyons, *Culture and Anarchy in Ireland, 1890–1939* (Oxford: Clarendon Press, 1979), 163.

3. J. J. Lee, *Ireland 1912–1985: Politics and Society* (Cambridge: Cambridge University Press, 1989), 237.

4. BBC News, "'Chuckle Brothers' Enjoy 100 Days," 27 August 2007, http://news.bbc.co.uk/1/hi/northern_ireland/6948406.stm, accessed 6 September 2011.

5. "Belfast Riots: A Setback for Area Barely Reshaped by Peace Process," *Guardian,* 22 June 2011; and "Belfast Riots Raise Tensions and Eyebrows," *Financial Times,* 22 June 2011.

6. See, for example, "Fear of an Uncertain Future for Romanian Immigrants," *Belfast Telegraph,* 18 June 2009, http://www.belfasttelegraph.co.uk/news/local-national/fear-of-an-uncertain-future-for-romanian-immigrants-14342527.html, accessed 6 September 2011; and "Islamic Centre Receives Far-Right Racist Threats," *Belfast Telegraph,* 1 July 2009, http://www.belfasttelegraph.co.uk/community-telegraph/south-belfast/news/islamic-centre-receives-farright-racist-threats-14375452.html, accessed 6 September 2011.

7. "Northern Ireland Has 'Culture of Intolerance,'" *Times,* 18 June 2011.

8. "Victoria Square Flats Open for Viewing," *Belfast Telegraph,* 26 April 2008, http://www.belfasttelegraph.co.uk/news/local-national/victoria-square-flats-open-for-viewing-13877027.html, accessed 6 September 2011; "Belfast's £5bn Titanic Quarter 'Where the City's Future Lies,'" *Belfast Telegraph,* 5 November 2010, http://www.belfasttelegraph.co.uk/news/local-national/northern-ireland/belfasts-5bn-titanic-quarter-development-lsquowhere-the-cityrsquos-future-liesrsquo-14995927.html, accessed 6 September 2011.

9. BBC News, "John Hume Hails Queen's Ireland Visit," 20 May 2011, http://www.bbc.co.uk/news/uk-northern-ireland-13467548, accessed 6 September 2011.

Notes on Methods and Literature

1. For detailed bibliographies on Irish history, see *Irish History Online,* http://www.irishhistoryonline.ie, accessed 25 June 2012; and K. Miller, *Irish History Bibliography,* http://history.missouri.edu/facultycourseinfo/IrishHistoryBibliog.pdf, accessed 25 June 2012.

2. R. F. Foster, *Modern Ireland 1600–1972* (Harmondsworth: Penguin, 1989); C. J. Beckett, *The Making of Modern Ireland, 1603–1923* (London: Faber, 1969); F. S. L. Lyons, *Ireland since the Famine* (London: Weidenfeld & Nicolson, 1971); M. Laffan, *The Partition of Ireland 1911–1925* (Dundalk: Dundalgan Press, 1983).

3. J. Mokyr, *Why Ireland Starved: A Quantitative and Analytical History of the Irish Economy, 1800–1850* (London: Allen & Unwin, 1985).

4. M. E. Daly, *The Slow Failure: Population Decline in Independent Ireland, 1920–1973* (Madison: University of Wisconsin Press, 2006); C. Ó Gráda, *Ireland: A New Economic History* (Oxford: Clarendon, 1994).

5. Lyons, *Ireland since the Famine.*

6. D. Massey, "Space-Time, 'Science' and the Relationship between Physical Geography and Human Geography," *Transactions of the Institute of British Geography* 24 (1999): 261–76.

7. E. A. Wrigley and R. S. Schofield, *Population History of England 1541–1871: A Reconstruction* (Cambridge, Mass.: Harvard University Press, 1981), 482.

8. L. Kennedy, P. S. Ell, E. M. Crawford, and L. A. Clarkson, *Mapping the Great Irish Famine: An Atlas of the Famine Years* (Dublin: Four Courts, 1999); J. Gleeson, R. Kitchin, B. Bartley, J. Driscoll, R. Foley, S. Fotheringham, and C. Lloyd, *The Atlas of the Island of Ireland: Mapping Social and Economic Change* (Kildare: All-Island Research Observatory, 2008).

9. B. M. S. Campbell and K. Bartley, *England on the Eve of the Black Death: An Atlas of Lay Lordship, Land and Wealth, 1300–49* (Manchester: Manchester University Press, 2006), 4.

10. See, for example, I. N. Gregory and R. G. Healey, "Historical GIS: Structuring, Mapping and Analysing the Geographies of the Past," *Progress in Human Geography* 31 (2007): 638–53; and A. K. Knowles, "GIS and History," in *Placing History: How Maps, Spatial Data, and GIS Are Changing Historical Scholarship,* ed. A. K. Knowles (Redlands, Calif.: ESRI, 2008), 1–25.

11. R. White, "What Is Spatial History," *The Spatial History Project* (2010), http://www.stanford.edu/group/spatialhistory/cgi-bin/site/pub.php?id=29, accessed 30 May 2012; D. J. Bodenhamer, "The Potential of Spatial Humanities," in *The Spatial Humanities: GIS and the Future of Humanities Scholarship,* ed. D. J. Bodenhamer, J. Corrigan, and T. M. Harris (Bloomington: Indiana University Press, 2010), 14–30; M. Dear, J. Ketchum, S. Luria, and D. Richardson, "Introducing the Geohumanities," in *GeoHumanities: Art, History, Text at the Edge of Place,* ed. M. Dear, J. Ketchum, S. Luria, and D. Richardson (Routledge: London, 2011).

12. *Historical GIS Research Network,* http://www.hgis.org.uk/bibliography.htm, accessed 30 May 2012, lists around 150 chapters but is by no means fully comprehensive.

13. B. M. S. Campbell, *English Seigniorial Agriculture, 1250–1450* (Cambridge: Cambridge University Press, 2000); B. Donahue, *The Great Meadow: Farmers and the Land in Colonial Concord* (New Haven, Conn.: Yale University Press, 2004); G. Cunfer, *On the Great Plains: Agriculture and Environment* (College Station: Texas A & M University Press, 2005); C. Gordon, *Mapping Decline: St. Louis and the Fate of the American City* (Philadelphia: University of Pennsylvania Press, 2008).

14. I. N. Gregory, C. Bennett, V. L. Gilham, and H. R. Southall, "The Great Britain Historical GIS: From Maps to Changing Human Geography," *Cartographic Journal* 39 (2002): 37–49; C. A. Fitch and S. Ruggles, "Building the National Historical Geographic Information System," *Historical Methods* 36 (2003): 41–51; M. De Moor and T. Wiedemann, "Reconstructing Belgian Territorial Units and Hierarchies: An Example from Belgium," *History and Computing* 13 (2001): 71–97. A. K. Knowles, ed., "Reports on National Historical GIS Projects," *Historical Geography* 33 (2005): 293–314 provides a review of many of these systems.

15. See G. Langran and N. R. Chrisman, "A Framework for Temporal Geographic Information," *Cartographica* 25 (1988): 1–14; G. Langran, *Time in Geographic Information Systems* (London: Taylor & Francis, 1992); and D. J. Peuquet, "It's about Time: A Conceptual Framework for the Representation of Temporal Dynamics in Geographic Information Systems," *Annals of the Association of American Geographers* 84 (1994): 441–61.

16. Langran and Chrisman, "Framework."

17. See, for example, D. J. Peuquet, *Representations of Space and Time* (London: Guildford, 2002).

18. A. H. Robinson, J. L. Morrison, P. C. Muehrcke, A. J. Kimerling, and S. C. Guptill, *Elements of Cartography,* 6th ed. (Chichester: John Wiley, 1995), esp. chap. 26.

19. See, for example, J. Langton, "Systems Approach to Change in Human Geography," *Progress in Geography* 4 (1972): 123–78; Massey, "Space-Time"; and D. Massey, *For Space* (London: Sage, 2005).

20. A. R. H. Baker, *History and Geography: Bridging the Divide* (Cambridge: Cambridge University Press, 2003); R. A. Butlin, *Historical Geography: Through the Gates of Space and Time* (New York: Routledge, 1993); R. A. Dodgshon, *Society in Space and Time: A Geographical Perspective on Change* (Cambridge: Cambridge University Press, 1998).

21. Campbell, *English Seigniorial Agriculture;* Cunfer, *On the Great Plains.*

22. Gordon, *Mapping Decline;* Donahue, *The Great Meadow.*

Index

Places are not indexed unless there is a substantive discussion concerning them.
Religious denominations are not indexed as they occur throughout the book.

Ian N. Gregory is Professor of Digital Humanities in the Department of History at Lancaster University.

Niall A. Cunningham is Research Associate at the ESRC Centre for Research on Socio-cultural Change (CRESC), University of Manchester.

C. D. Lloyd is Senior Lecturer in the Department of Geography and Planning, School of Environmental Sciences, University of Liverpool.

Ian G. Shuttleworth is Senior Lecturer in the School of Geography, Archaeology, and Palaeoecology at the Queen's University Belfast.

Paul S. Ell is Director of the Centre for Data Digitisation and Analysis in the School of Geography, Archaeology, and Palaeoecology at the Queen's University Belfast.